LASER 2004

LASER 2004

Proceedings of the 6th International Workshop on Application of Lasers in Atomic Nuclei Research (LASER 2004) held in Poznań, Poland, 24–27 May 2004

Edited by

Z. BŁASZCZAK

Adam Mickiewicz University, Poznań, Poland

B. MARKOV

Joint Institute for Nuclear Research, Dubna, Russia

and

K. MARINOVA

Joint Institute for Nuclear Research, Dubna, Russia

Reprinted from *Hyperfine Interactions*
Volume 162, Nos. 1–4 (2005)

Springer

A C.I.P. Catalogue record for this book is available from the Library of Congress.

ISBN 3-642-44695-7

Published by Springer
P.O. Box 990, 3300 AZ Dordrecht, The Netherlands

Printed on acid-free paper

Table of Contents

ORGANIZING COMMITTEE

Chairmen: Z. Błaszczak
B. Markov
Scientific secretary: K. Marinova
Members: J. Billows, S. Choinacki, J. Dembczyński,
Yu. P. Gangrsky, W. Nawrocik
Yu. E. Penionzhkevich, S. G. Zemlyanoi

INTERNATIONAL ADVISORY COMMITTEE

H. Backe (*Mainz, Germany*), F. Le Blanc (*Orsay, France*), P. van Duppen (*Leuven, Belgium*), V. Fedoseev (*CERN*), T. T. Inamura (*RIKEN, Japan*), H.-J. Kluge (*GSI, Germany*), J. Kownacki (*Warsaw, Poland*), V.I. Mishin (*Troitsk, Russia*), R. Neugart (*Mainz, Germany*), Yu. Oganessian (*JINR, Russia*)

Organized by

Faculty of Physics, Adam Mickiewicz University (AMU), Poznań, Poland
Flerov Laboratory of Nuclear Reactions (FLNR), JINR, Dubna

Hyperfine Interactions (2005) 162:1–2
DOI 10.1007/s10751-005-9201-5

Preface

The VI International Workshop on the Application of Lasers in Atomic Nuclei Research entitled "Laser Methods in the Study of Nuclei, Atoms and Molecules," took place in Poznań, May 24th–27th, 2004. It was organised within the international collaboration between the Flerov Laboratory of Nuclear Reactions, JINR, Dubna, Russia and the Faculty of Physics, Adam Mickiewicz University (AMU), Poznań, Poland. Starting from 1990, the Workshops have been organised once every two to three years. The aim of the regular meetings is to report recent achievements, check which of the expectations and hopes have come true, to discuss the perspectives of further work in the field and the mutual impact on other branches of science. Each workshop has been focused on a somewhat different main theme. The earlier workshops were concentrated on:

- Laser spectroscopy of atomic nuclei, Dubna, December 18–20, 1990;
- Charge and nuclear radii of exotic nuclei, Poznań, May 29–31, 1995;
- Hyperfine structure and nuclear moments of exotic nuclei by laser spectroscopy, Poznań, February 3–5, 1997;
- Laser spectroscopy on beams of radioactive nuclei, Poznań, May 24–27, 1999;
- Prospects for development of laser methods in the study of nuclear matter, Poznań, May 28–31, 2001.

At the opening ceremony we honoured the memory of Professor Sylwester Chojnacki, our colleague and member of the Organising Committee, who has passed away in the beginning of May this year. Professor Chojnacki will be long remembered by us for this great devotion to science and great heart.

Today, at many scientific centres, laser methods find increasing area of application. The VI International Workshop was dedicated especially to the laser methods in the study of nuclei, atoms and molecules. The scientific programme covered a broad list of topics highlighting recent research in the following fields: contemporary laser spectroscopic techniques of different types; laser spectroscopy and nuclear structure of exotic nuclei, fission fragments and transuranium elements; hyperfine anomaly and nuclear magnetisation radii; detection of trace elements by laser spectroscopy and different applications of lasers and nuclear technique in life-science.

The most important scientific centres working in the area of laser application in nuclear physics research include: Argonne National Laboratory, CERN, GSI

(Darmstadt), IGISOL (Jyväskylä), JINR (Dubna), RIKEN (Wako), TRIUMF (Vancouver) as well as the Universities of Darmstadt, Leuven, McGill, Mainz, Manchester, Munich. The total number of participants representing these centres was 48 from 12 countries; participation of young physicists and PhD students was particularly encouraged. The number of participants from foreign countries as well as of the countries represented hit a record high since the first workshop was organised. The organisers clearly intended that the workshop should be one of the European meetings of physics in future. In fact, the organisers set up the international advisory committee for the first time. The proceedings have been for the first time published by a prominent journal.

At the Conference forty talks, including 14 invited, were presented. The contributions of the young participants made about 20%. The talks were in generally devoted to the new generation of experimental methods and indicated that the main interest switched from a systematic exploration of larger regions of the chart of nuclides to the poorly investigated ‘blank fields’ and the applications. A detailed report of Prof Inamura (RIKEN) on the workshop progress, contributions and achievements has already appeared in “Comments on Atomic, Molecular and Optical Physics” (Physica Scripta. Vol. 71, C1–4, 2005)

The workshop was completed with a very exciting discussion of perspectives for future investigations and the need for development of collaborations. The organisers hope that the workshop will help in the choice of the most interesting and promising experiments and will promote a joint use of the experimental possibilities of different institutes, helping initiate new collaborations.

We would like to thank all participants for coming to Poznań and for their contributions. We do appreciate the efforts of all the members of the Local and International Committee for making this Workshop a success. Special thanks are due to Prof J. Kluge (GSI, Darmstadt) for the opportunity of publishing the workshop proceedings in Hyperfine Interactions and for his huge editorial work.

The organisers wish to acknowledge the financial support within the JINR-Poland grant, and the support from the Rectors of AMU and the Technical University of Poznań.

We are much grateful to everybody who contributed to organising the Workshop.

Zdzisław Błaszczak
Chairman

Boris Markov
Co-chairman

Krassimira Marinova
Scientific secretary

Hyperfine Interactions (2005) 162:3–14
DOI 10.1007/s10751-005-9209-x

Laser Spectroscopic Investigation of the Element Fermium ($Z = 100$)

H. BACKE[1,*], A. DRETZKE[1], ST. FRITZSCHE[3], R. G. HAIRE[4], P. KUNZ[1], W. LAUTH[1], M. SEWTZ[1] and N. TRAUTMANN[2]
[1]*Institut für Kernphysik der Universität, D-55099 Mainz, Germany; e-mail: backe@kph.uni-mainz.de*
[2]*Institut für Kernchemie der Universität, D-55099 Mainz, Germany*
[3]*Fachbereich Physik, Universität Kassel, D-34109 Kassel, Germany*
[4]*Oak Ridge National Laboratory, Oak Ridge, TN 37831, USA*

Abstract. The level scheme and hyperfine structure of the element fermium ($Z = 100$) has been investigated with the method of resonance ionization spectroscopy in a buffer gas cell. The experiments were carried out on a 46 pg sample of the isotope ^{255}Fm with a half-life time of 20.1 h, produced in the high flux nuclear reactor of the ORNL, Oak Ridge, USA. A wave number scan from 27,100 to 28,400 cm^{-1} was carried through to search for three levels with large Einstein-coefficients, as predicted by *ab initio* Multi-Configuration-Dirac-Fock (MCDF) calculations, and five new levels were found. In addition, the two known levels at wave numbers (25,099.8 ± 0.2) and (25,111.8 ± 0.2) cm^{-1} were studied with a laser band width of 1.5 GHz and hyperfine broadenings were observed.

Key Words: atomic level scheme, heavy element fermium, hyperfine structure, resonance ionization.

1. Introduction

One of the most fascinating studies of the heaviest actinides and the trans-actinides concerns the influence of increasing relativistic effects on the valence-electron configuration of the atoms and its consequence on chemical behavior. Relativistic effects gain importance for the heaviest elements in the region of $Z = 100$ with the consequence that simple extrapolation of systematics – e.g., the electron affinity – may not lead to reliable predictions any more. These relativistic effects are caused – roughly speaking – by a shrinking of the wave functions of s- and $p_{1/2}$- electrons. Inner shell electrons influence indirectly via the shielding of the nuclear potential the valence electrons as well.

The valence electron configuration can be predicted by *ab initio* Multi-Configuration-Dirac-Fock (MCDF) calculations and other methods [1–3], but these calculations are complicated and have limited accuracy. The results,

* Author for correspondence.

therefore, need to be checked by experiments. A critical test of such MCDF calculations on electron configurations of *trans*-einsteinium elements would be to compare properties such as first ionization potentials and the excitation schemes, including spins and lifetimes of the levels, with the experimental values. The experimental methods must be sensitive enough and be applicable to *trans*-einsteinium elements at low production rates (less than 1 atom per second), for short live times of the elements, and for no atomic excitation schemes known. Such methods are based on laser spectroscopy and have been applied for the element fermium ($Z = 100$) [4].

The atomic level structure of the element fermium ($Z = 100$) was recently investigated for the first time with the Ion-Guide detected Resonance Ionization Spectroscopy (IGRIS) method [4]. This method combines Resonance Ionization Spectroscopy (RIS) which is sensitive to the nuclear charge number Z and the Ion-Guide Quadrupole Mass Separation (IGQMS) which is sensitive to the mass number A [5]. A sample of only $2.7 \cdot 10^{10}$ atoms of the isotope ^{255}Fm with a half-life of 20.1 h was available. It has been demonstrated with this experiment that a successful level search is possible with our laser spectroscopic technique even if the number of atoms is rather low and no experimental information on the level scheme is available. The two levels which were found have been predicted by MCDF calculations. Partial lifetimes were determined from saturation characteristics and term assignments were proposed. The agreement between experiment and calculations is striking. Rather surprisingly the results were noticed in [6]. To get insight into the accuracy of such *ab initio* MCDF predictions, further levels for the element fermium were sought for. Three $5f^{12}7s7p$ levels with large Einstein coefficients A_{ik} in the order of $2–4 \cdot 10^8$ have been predicted by MCDF calculations in the energy range between 27,394 and 27,802 cm^{-1}. In addition, we investigated the hyperfine splitting of the two already observed levels at energies of $(25{,}099.8 \pm 0.2)$ and $(25{,}111.8 \pm 0.2)$ cm^{-1} with a laser band width of 1.5 GHz aiming for a determination of the hyperfine structure constants A and B. Both experiments are described in the following section.

2. Experimental

The experimental setup is shown in Figure 1. As opposed to conventional IGISOL apparatus where the gas jet alone transports the ions out of the cell, in this cell they drift along electric field lines created by a suitable electrode system. Shortly before the ions enter the nozzle they are decoupled from the electrical field and flushed out by the gas jet. With a nozzle diameter of 1 mm and argon at a pressure of 35 mbar as buffer gas, pumps with relatively low pumping speeds can be used (TMP I–IV). Thus, the apparatus remains rather compact. With the aid of a segmented quadrupole ion guide structure and skimmers, the ions are separated from the buffer gas, mass selected in a quadrupole mass spectrometer (Balzers QMG 311) and identified by a channeltron detector.

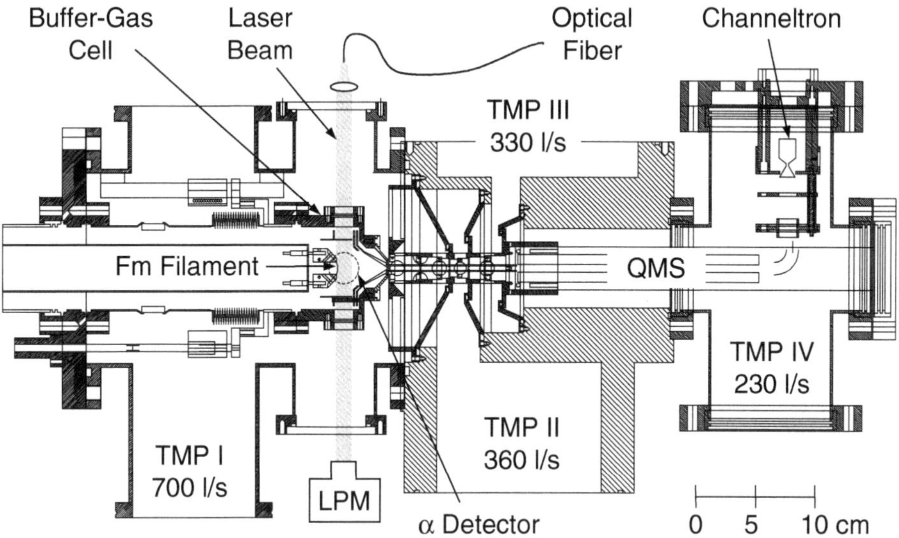

Figure 1. Experimental setup. The four sections are evacuated by Turbo Molecular Pumps (TMP) with relatively low pumping speeds as indicated. *QMS* is the Quadrupole Mass Spectrometer and *LPM* the Laser Power Meter.

The isotope ^{255}Fm with a half-life time $t_{1/2} = 20.1$ h was produced at ORNL in Oak Ridge, USA. Starting with the isotope ^{246}Cm, the isotope ^{255}Es was bred in the high-flux isotope reactor (HFIR) at a neutron flux of $2.6 \cdot 10^{15}$ n/(cm^2 · s) by successive neutron captures and β^-decays. A sample of 4 ng of the isotope ^{255}Fm was chemically extracted and air shipped to Germany. In addition, a sample of 1 pg ^{257}Fm with a half-life time $t_{1/2} = 100.5$ d was produced. Half of this amount was also shipped to Germany for our experiments. At the Institute of Nuclear Chemistry at the University of Mainz ^{255}Fm atoms were electro-deposited onto a tantalum filament and covered with approximately 1 μm of titanium. Three filaments were produced, two with numbers of $3 \cdot 10^{10}$ and $8 \cdot 10^{10}$ atoms of ^{255}Fm, and one with $3 \cdot 10^{8}$ atoms ($\triangleq 0.13$ pg) of ^{257}Fm.

The fermium atoms, evaporated from the electrically heated filament, were resonantly ionized employing a pulsed laser system. The laser system consisted of an excimer laser (EMG104 MSC) which runs on XeF at a wavelength of 351/353 nm with a repetition rate of 200 Hz. It provides 50 mJ pulses with a duration of 15 ns, and pumps a tunable dye laser (Lambda, FL2002) with a spectral line width of 6–10 GHz. A second excimer laser which runs on XeCl at a wavelength of 308 nm with a repetition rate of 100 Hz provides 300 mJ pulses with a duration of 20 ns and pumps two more tunable dye lasers (Lambda, FL2001 and FL3002). The wavelength was monitored using a wavemeter (Atos, LM007). The performance of the laser systems was checked during the course of the experiment on the element erbium in a reference cell.

Table I. Results of MCDF calculations. Listed are wave number $\overline{\nu}$, total angular momentum quantum number J, transition rate A_{ki}, configuration in the LS coupling scheme with largest expansion coefficient c, term assignment and $|c|^2$. Taken from [4, 7].

No.	$\bar{\nu}$ [cm^{-1}]	J	A_{ki} [1/s]	Config.	Term	$\|c\|^2$
0	0	6	0	$5f^{12}7s^2$	$^3H_6^e$	0.96
1	25,226	6	$1.89 \cdot 10^6$	$5f^{12}7s7p$	$^5I_6^o$	0.46
2	25,471	5	$1.28 \cdot 10^6$	$5f^{12}7s7p$	$^5G_5^o$	0.34
3	27,394	6	$2.43 \cdot 10^8$	$5f^{12}7s7p$	$^3H_6^o$	0.62
4	27,633	5	$1.98 \cdot 10^8$	$5f^{12}7s7p$	$^3G_5^o$	0.60
5	27,802	7	$3.67 \cdot 10^8$	$5f^{12}7s7p$	$^3I_7^o$	0.66
6	28,540	5	$2.82 \cdot 10^5$			
7	29,359	5	$3.58 \cdot 10^7$			

3. Investigation of the level structure

Ab initio MCDF calculations predict three $5f^{12}7s7p$ levels around 27,500 cm^{-1} with large transition rates, see level numbers 3–5 in Table I. To search for these levels the laser systems described in the last subsection were employed. The lasers were synchronized and operated at 50 Hz repetition rate. Two dye lasers scanned the wavelength regions 25,970 cm^{-1} $\leq \overline{\nu}_1 \leq$ 26,300 cm^{-1} and 27,100 cm^{-1} $\leq \overline{\nu}_1 \leq$ 28,400 cm^{-1}. The second excitation step into the continuum was accomplished by light with a wavelength of 351/353 nm, delivered by the excimer laser running on XeF. At the scan in the wavelength region 25,970 cm^{-1} $\leq \overline{\nu}_1 \leq$ 26,300 cm^{-1} no transitions were observed. The results of the second dye laser scan are shown in Figure 2. In Table II the level energies together with line widths are listed. For energy calibration seven well-known reference transitions in erbium in the energy range between 27,582.017 and 28,129.803 cm^{-1} have been used.

From the viewpoint of the accuracy of the *ab initio* MCDF calculations of $\pm$2,400 cm^{-1} all five observed lines are candidates for the three $5f^{12}7s7p$ levels with large A_{ki}. Therefore, it must be concluded that the estimated accuracy of the MCDF calculations seems to be realistic. The very good agreement of the experimental level energy and the MCDF calculations as reported in [4] for the previously observed levels at 25,226 and 25,471 cm^{-1} are probably of accidental nature.

It can be seen from Table I that the levels with large A_{ki} could be identified by a measurement of the ground state transition rates. The total transition rate, including radiationless quenching transitions in buffer gas collisions, can be determined from the decrease of the ionization rate if the time delay between populating and ionizing laser pulse is increased. For such a measurement the time delay t_d between the dye laser pulse and the ionizing excimer laser pulse was varied. From these measurements, see Figure 3, only an upper limit of the life-

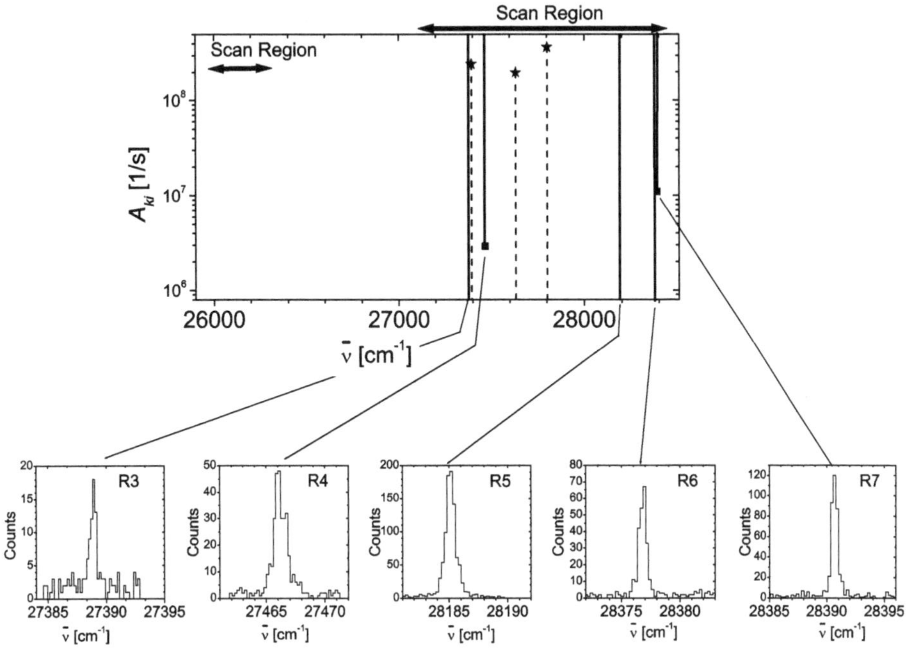

Figure 2. Results of MCDF calculations with transition rates A_{ki} (★-*dashed*) and experimental observed transitions in the energy range from 27,100 to 28,400 cm^{-1} (■-*black line*). In the *lower part* dye laser scans of the transitions are shown.

Table II. Experimental results of the resonance ionization spectroscopy at ^{255}Fm. Quoted are also line width (FWHM) and transition rate A_{ki}

No.	$\bar{\nu}$ [cm^{-1}]	FWHM [cm^{-1}]	A_{ki} [1/s]	Ref.
R1	25,099.80 ± 0.2	–	$(3.4 \pm 0.8) \cdot 10^6$	[4]
R2	25,111.80 ± 0.2	–	$(3.5 \pm 0.7) \cdot 10^6$	[4]
R3	27,389 ± 1.5	0.85 ± 0.16		This work
R4	27,466 ± 1.5	1.34 ± 0.09	$\geq 2.9 \cdot 10^6$	This work
R5	28,185 ± 1.5	1.08 ± 0.05		This work
R6	28,377 ± 1.5	0.75 ± 0.05		This work
R7	28,391 ± 1.5	0.61 ± 0.03	$\geq 1.1 \cdot 10^7$	This work

time of $\tau_i \leq 345$ ns of the transition at 27,466.0 cm^{-1} (R4) and a lifetime of $\tau_i \leq 90$ ns of the transition at 28,391.0 cm^{-1} (R7) could be estimated because either the statistics were poor or the point density insufficient. If quenching and branching transitions into lower levels can be neglected, estimated lower limits $A_{ki} = 1/\tau_{ki} = 1/\tau_i \geq 2.9 \cdot 10^6$/s and $A_{ki} \geq 1.1 \cdot 10^7$ result for transitions R4 and R7, respectively. A comparison with Table I allows only to conclude that transition R4 and R7 cannot be assigned to the transition from level 6. Unfortu-

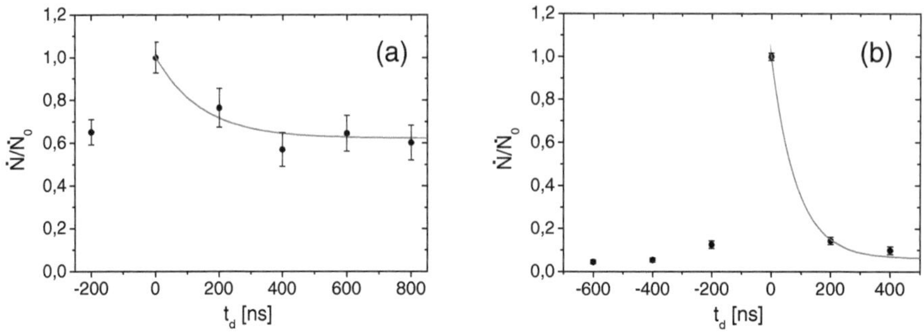

Figure 3. Resonance ionization signal as function of the time delay t_d between first and second laser pulse for the transition at 27,466 cm^{-1} (a) and 28,391 cm^{-1} (b). The *fit curves* allow only estimates of upper limits of the lifetime $\tau_i = (143 + 202)$ ns = 345 ns (a) and $\tau_i = (85 + 10)$ ns = 95 ns (b). These upper limits were obtained by increasing the fit value by its standard deviation.

nately, the atom number of the ^{255}Fm sample was not sufficient to perform better measurements.

A question arises whether some information on the transition rate may be obtained from the line widths of the transitions which are also quoted in Table II. However, since no saturation characteristics as a function of the pulse energy flux of the dye laser populating the levels could be measured, reliable results cannot be extracted from these values.

4. Hyperfine splitting

Part of the fermium sample was used to investigate the hyperfine splitting of the previously observed transitions at 25,099.8 and 25,111.8 cm^{-1} [4]. The bandwidth of the dye laser was reduced to 1.5 GHz (FWHM), corresponding to 0.05 cm^{-1}, by means of an intracavity etalon. To avoid power broadening a low pulse energy flux of $\Phi_1 = 0.86 \cdot 10^{14}$ photons/(cm^2 · pulse) was coupled into the optical cell. The results of the measurements are depicted in Figure 4 by the error bars.

The atomic ground state of fermium has a total angular momentum quantum number $J_i = 6$ as has been concluded from [8]. MCDF calculations suggest that one of the excited levels has a total angular momentum quantum number $J_f = 6$, the other one $J_f = 5$, see Table I. In [4] it was not possible to assign quantum numbers to the observed levels. Consequently, both assignments were assumed in the analysis of the hyperfine structure measurements. With a nuclear spin of $I = 7/2$ [9], a total of 22 and 21 hyperfine structure components result for $J_f = 6$ and $J_f = 5$, respectively. With this information least square fits were performed even though the hyperfine structure components could not be resolved in the measurements. In the fit procedure the hyperfine structure constants for

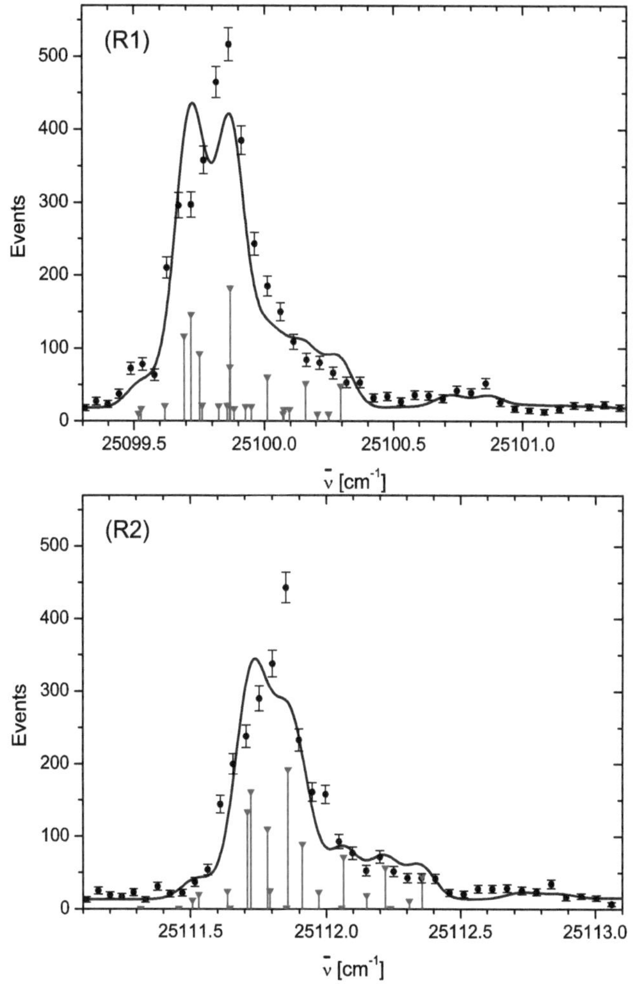

Figure 4. Wave number scans with a spectral bandwidth of $\Delta\nu$ = 1.5 GHz (FWHM) for the transitions to the levels at 25,099.8 cm^{-1} (R1) and 25,111.8 cm^{-1} (R2). Measurements are shown by the *error bars*. Due to the free spectral range of the etalon (30 GHz) the *lines* appear shifted by 30 GHz once more with reduced intensity. The *full line* shows the best fit with relative intensities according to Equation (1). The corresponding hyperfine structure components and their relative intensities are indicated by symbols (▼).

ground and excited states A_i, A_f and B_i, B_f, the transition energy, the amplitude and the background were kept as free parameters. The accuracy could be improved by a simultaneous fit using the same ground state parameters for both transitions. The laser bandwidth of 1.5 GHz and a Doppler broadening of 0.5 GHz were incorporated in the fit procedure as Gaussians. The natural Lorentzian line widths of the individual hyperfine structure components were assumed to be negligible compared to the Gaussians. A possible power broadening was not taken into account. The relative intensities of the hyperfine structure components are assumed to be given by [10]

$$S(F_i \rightarrow F_f) = \frac{(2F_f+1)(2F_i+1)}{2I+1}\begin{Bmatrix} J_f & F_f & I \\ F_i & J_i & 1 \end{Bmatrix}^2. \tag{1}$$

Here $\{\because\}$ is the 6*j*-symbol, and F_i and F_f are the total angular momentum quantum numbers of the ground and excited states, respectively.

Results of the best fits with the assumptions of the hyperfine transitions according to Equation (1) are also shown in Figure 4. The best agreement resulted with the assumption of $J_i = 6 \rightarrow J_f = 6$ for transition R1 and $J_i = 6 \rightarrow J_f = 5$ for transition R2, but a definite assignment of J_f was not possible. The hyperfine structure constants resulting from this analysis are: $A_g \simeq -0.32$ GHz and $B_g \simeq -22$ GHz for the ground state; $A_e^{(1)} \simeq -0.53$ GHz and $B_e^{(1)} \simeq -2.9$ GHz for the excited state R1; $A_e^{(2)} \simeq -0.69$ GHz and $B_e^{(2)} \simeq -1.7$ GHz for the excited state R2.

The reason for the rather poor fit, see Figure 4, might be that the fit model does not take into account the saturation of transitions and pumping processes between hyperfine structure (hfs) components. Even a partial saturation would already alter the relative intensities according to Equation (1). In addition, in case of an overlap of several hfs levels in ground and excited state with the laser band width, a population redistribution of the hfs levels may occur. An upward transition from one certain ground state hfs level may be followed by a stimulated downward transition into another ground state level. To calculate the occupation numbers M_i of the hfs levels of the excited state, which are directly proportional to the resonance ionization probability, the following rate equations were solved numerically:

$$\frac{dN_i}{dt} = W_{i,i-1}M_{i-1} + W_{i,i}M_i + W_{i,i+1}M_{i+1} - \left(W_{i,i-1} + W_{i,i} + W_{i,i+1}\right)N_i \tag{2}$$

$$\frac{dM_i}{dt} = W_{i-1,i}N_{i-1} + W_{i,i}N_i + W_{i+1,i}N_{i+1} - \left(W_{i-1,i} + W_{i,i} + W_{i+1,i}\right)M_i \tag{3}$$

Here N_i are the occupation numbers of the hfs components of the ground state and W_{ij} the transition rates between the ground and excited hfs levels. The latter are given by the equation

$$W_{i,j} = W(\omega_L, F_i, F_j) = \frac{\lambda_0^2}{4} A_{kl} \; \frac{I_0}{\hbar\omega_0} S(F_i \rightarrow F_j) \times \int_{-\infty}^{\infty} I_L(\omega - \omega_L, \omega_0, \sigma_L) I_S(\omega, \omega_0, \Gamma, \sigma_D)\, d\omega. \tag{4}$$

where λ_0 and ω_0 are the resonance wavelength and frequency, A_{kl} is the spontaneous transition rate (see Table II), I_0 is the laser intensity and ω_L the laser

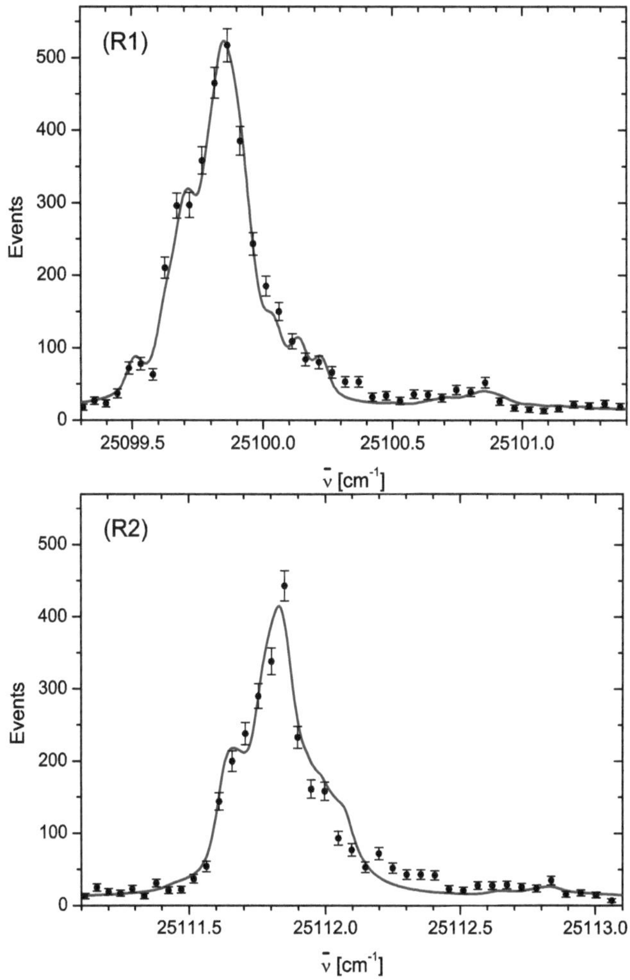

Figure 5. Fit of the experimental data with a rate equation minimization procedure which takes saturation effects and pumping processes into account.

frequency. The integral describes the overall spectral function of the transition. It consists of a spectral function of the laser I_L with a Gaussian profile of width σ_L and a convolution of the natural line width Γ and a Doppler profile of width σ_D. Since for ^{255}Fm $J_i = 6$, $J_f = 5,6,7$, and $I = 7/2$ the number of hyperfine levels in the ground and excited state is always eight, 16 coupled equations must be solved.

The rate equations were solved by integration over the laser pulse width of 20 ns with the assumption of a laser pulse intensity of 133 $\mu J/cm^2$. The best fits are shown in Figure 5. They reproduce the line profiles much better than the fits based on Equation (1) (Figure 4), but a definite assignment of the angular momentum quantum numbers J_f was not possible. Several assumptions were tested. While fits with the $J_i = 6 \rightarrow J_f = 7$ configuration did not yield satisfying results, fits with the $J_i = 6 \rightarrow J_f = 5, 6$ configurations produced similar results within the margin of error. The hyperfine structure constants associated with the example in Figure 5 are $A_g \simeq -0.76$ GHz, $B_g \simeq -8.7$ GHz for the common ground state ($J_i = 6$) and $A_e^{(1)} \simeq -0.66$ GHz, $B_e^{(1)} \simeq -14.8$ GHz for the excited state R1 ($J_f = 6$) and $A_e^{(2)} \simeq -0.73$ GHz, $B_e^{(2)} \simeq 0.39$ GHz for the excited state R2 ($J_f = 5$).

It should be stressed that the A and B factors of ground and excited states strongly correlated with each other. Quite reasonable fits were also obtained with hyperfine constants up to a factor of 1.5 larger ($B_e^{(2)}$ of R2 varied within up to a factor of 20). With the exception of the latter, we estimated the errors of the hyperfine constants to be in the order of about 50%. Despite of these large errors it might be instructive to compare our results with the hyperfine constants of neighboring actinide and homolog lanthanide isotopes with the same nuclear spin $I = 7/2$. The homologous elements holmium and einsteinium have $4f^{11}6s^2\ {}^4I_{15/2}$ and $5f^{11}6s^2\ {}^4I_{15/2}$ terms for their ground states, respectively. The magnetic moments μ_I, A factors, the spectroscopic quadrupole moments Q_s, and B factors for ^{165}Ho [11] and for ^{253}Es [12] are $\mu_I^{(Ho)} = 4.132\ \mu_n$, $A^{(Ho)} = 0.80058$ GHz, $Q_s^{(Ho)} = 3.58$ b, $B^{(Ho)} = -1.6681$ GHz, and $\mu_I^{(Es)} = 4.1\ \mu_n$, $A^{(Es)} = 0.817$ GHz, $Q_s^{(Es)} = 6.7(8)$ b, $B^{(Es)} = -4.316$ GHz, respectively. From these numbers quite similar magnetic hyperfine fields $\langle H(0)\rangle_{Ho} = I \cdot JA/\mu_I = 5.09$ GHz μ_n and $\langle H(0)\rangle_{Es} = 5.23$ GHz$/\mu_n$ can be calculated for holmium and einsteinium, respectively, the corresponding numbers for the electric field gradients are $\langle \varphi_{jj}(0)\rangle_{Ho} = B/(eQ_s) = -0.466$ GHz/eb and $\langle \varphi_{jj}(0)\rangle_{Es} = -0.644$ GHz/eb. It might be tempting to speculate that the ratios $\langle H(0)\rangle_{Es}/\langle H(0)\rangle_{Ho}$ and $\langle \varphi_{jj}(0)\rangle_{Es}/\langle \varphi_{jj}(0)\rangle_{Ho}$ do not substantially change for the neighboring homologue elements erbium and fermium since in the ground states $4f^{12}6s^2\ {}^3H_6$ and $5f^{12}6s^2\ {}^3H_6$ only one f electron is added, and since both nuclear spins for the isotopes ^{167}Er and ^{255}Fm are again $I = 7/2$. Then, from the values $\mu_I^{(Er)} = -0.56385\ \mu_n$, $A^{(Er)} = -0.120$ GHz, $Q_s^{(Er)} = 3.565$ b, $B^{(Er)} = -4.552486$ GHz for ^{167}Er [13] we obtain for ^{255}Fm hyperfine fields $\langle H(0)\rangle_{Fm} = 4.61$ GHz$/\mu_n$ and $\langle \varphi_{jj}(0)\rangle_{Fm} = -1.765$ GHz/eb. From the above quoted fit values for the A and B factors of the ^{255}Fm ground state we derive a magnetic moment $\mu_I^{(Fm)} = -3.46\ \mu_n$ and a spectroscopic quadrupole moment $Q_s^{(Fm)} = 4.93$ b, the latter corresponding to an intrinsic quadrupole moment $Q_0^{(Fm)} = (I+1)(2I+3)/(I(2I-1))Q_s^{(Fm)} = 10.6$ b. The magnetic moment seems to have an unreasonable large negative value, exceeding the Schmidt value of $-1.9\ \mu_n$. However, the intrinsic quadrupole moment is in accord with the value of about 12 b found for well deformed nuclei in this domain of the nuclear table [9].

Finally, we tried to determine the isotope shift between ^{255}Fm and ^{257}Fm for the 25,099.8 and 25,111.8 cm^{-1} transitions. However, a measurement could not be performed since the 1 pg sample of ^{257}Fm turned out to be not sufficient.

5. Conclusions

The atomic level structure of the element fermium ($Z = 100$) was investigated with the Ion-Guide detected Resonance Ionization Spectroscopy (IGRIS) method in a buffer gas cell.

In a former experiment two levels at 25,100 and 25,112 cm^{-1} were found which agree with predictions of *ab initio* Multi-Configuration-Dirac-Fock (MCDF) calculations with a striking good accuracy of less than 371 cm^{-1}, at an estimated accuracy of the MCDF calculations of ±2,400 cm^{-1} [4]. To get further insight into the predicting power of such MCDF calculations, a search for three predicted levels at energies of 27,394, 27,633, and 27,802 cm^{-1} with large ground state transition rates of about $A_{ki} = (2 - 4) \cdot 10^8$/s has been carried through.

Five transitions were found. From the viewpoint of the accuracy of the *ab initio* MCDF calculations of ±2,400 cm^{-1} all five observed lines are candidates for the three predicted levels with large A_{ki}. From this fact it must be concluded that the estimated accuracy of the MCDF calculations seems to be realistic. The very good agreement for the previously observed levels and the MCDF calculations, as reported in [4], is probably of accidental nature. The identification of the predicted states would be possible by a measurement of the ground state transition rates A_{ki}. Unfortunately, the limited number of ^{255}Fm atoms allowed only the quotation of lower limits of the transition rates for two transitions.

A part of the fermium sample was used to investigate the hyperfine splitting of the previously observed transitions at 25,099.8 and 25,111.8 cm^{-1}. The bandwidth of the scanning laser was reduced to 1.5 GHz (FWHM) by means of an intracavity etalon. Hyperfine components could not be resolved, however, significantly broadened structures were observed. Hyperfine structure constants were deduced by a best fit of a model structure to the experimental data, which takes into account saturation and optical pumping effects.

Acknowledgements

We would like to thank the R.E.D.C. facility, ORNL, for their efforts in separating and processing the Fm isotopes from the target products and the Division of Chemical Sciences, Office of Basic Energy Research, U.S. Department of Energy, for making the Fm material available through the trans-plutonium element production program at ORNL. We also thank K. Eberhardt, G. Passler and P. Thörle for their active support during the experiment. This work has been supported by the Bundesministerium für Bildung und Forschung under contract No. 06 MZ 169I and European Union EU under contract NIPNET HPRI-CT-2001-50034.

References

1. Fricke B., Johnson E. and Rivera G. M., *Radiochim. Acta* **62** (1993), 17.
2. Martin W. C. and Sugar J., *Phys. Rev.* **A 53** (1996), 1911.
3. Eliav E., Kaldor U. and Ishikawa Y., *Phys. Rev. Lett.* **74** (1995), 1079.
4. Sewtz M., Backe H., Dretzke A., Kube G., Lauth W., Schwamb P., Eberhardt K., Grüning C., Thörle P., Trautmann N., Kunz P., Lassen J., Passler G., Dong C.Z., Fritzsche S. and Haire R.G., *Phys. Rev. Lett.* **90**(16) (2003), 163001–163002.
5. Backe H., Eberhardt K., Feldmann R., Hies M., Kunz H., Lauth W., Martin R., Schöpe H., Schwamb P., Sewtz M., Thörle P., Trautmann N. and Zauner S., *Nucl. Instr. Meth.* **B 126** (1997), 406.
6. Baker O., *Phys. Rev. Focus*, 25. April 2003.
7. Fritzsche S., *Proceedings of the International Workshop on Ion Chemical Aspects in Ion Guide Systems*, University of Mainz, Sept. 1999.
8. Stanton H. E., Goodman L. S., Diamond H. and Fred M. S., *Phys. Rev. A* **4**(2) (1971), 473.
9. Lederer C. M. and Shirley V. S., *Table of Isotopes*, 7th edn., Wiley, New York, 1987.
10. Sobelman I. I., *Atomic Spectra and Radiative Transitions*, Springer, Berlin Heidelberg New York, 1992.
11. Merzyn G., Penselin S. and Wolber G., *Z. Physik*, **252** (1972), 412–420.
12. Blaise J. and Wyart J.-F., *Selected Constants, Energy Levels and Atomic Spectra of Actinides*, Vol. 20, Tables Internationales de Constantes, Université P. et M. Curie, Paris, 1992.
13. Childs W. J., Goodman L. S. and Pfeufer V., *Phys. Rev.* **A 28**(6) (1983), 3402

Hyperfine Interactions (2005) 162:15–27
DOI 10.1007/s10751-005-9204-2

Ionization Scheme Development at the ISOLDE RILIS

V. N. FEDOSSEEV[1], B. A. MARSH[1,2,*], D. V. FEDOROV[3], U. KÖSTER[1] and E. TENGBORN[4]
[1]*CERN, 1211, Geneva-23, Switzerland*
[2]*The University of Manchester, Manchester M13 9PL, UK; e-mail: Bruce.Marsh@cern.ch*
[3]*Petersburg Nuclear Physics Institute, 188350, Gatchina, Russia*
[4]*Chalmers University of Technology, Göteborg 41296, Sweden*

Abstract. The resonance ionization laser ion source (RILIS) of the ISOLDE on-line isotope separation facility is based on the method of laser step-wise resonance ionization of atoms in a hot metal cavity. The atomic selectivity of the RILIS complements the mass selection process of the ISOLDE separator magnets to provide beams of a chosen isotope with greatly reduced isobaric contamination. Using a system of dye lasers pumped by copper vapour lasers, ion beams of 24 elements have been generated at ISOLDE with ionization efficiencies in the range of 0.5–15%. As part of the ongoing RILIS development off-line resonance ionization spectroscopy studies carried out in 2003 and 2004 have determined the optimal three-step ionization schemes for scandium, antimony, dysprosium and yttrium.

Key Words: radioactive ion beams, resonance ionization laser ion source, Dy, Sb, Sc, Y.

1. Introduction

ISOLDE is an isotope separator on-line (ISOL) type radioactive ion beam facility. Radioactive atoms are produced in a thick target during its bombardment by a high-energy proton beam. Isotope separation of the ionized reaction products takes place as the ion beam passes through a magnetic mass separator. This process alone does not provide a chemically pure beam since many isobars may be present at the chosen mass. Thus, an additional separation between nuclides with different proton number Z is favored for many applications of the radioactive ion beams (RIBs). This can be performed by chemical methods, using the different chemical behavior of different elements. Alternatively, an atomic physics technique, step-wise resonance photo-ionization, can be used. At the ISOLDE facility, the resonance ionization laser ion source (RILIS) exploits the unique electronic structure of different atomic species to provide a rapid,

* Author for correspondence.

efficient and Z-selective ionization process [1, 2]. In principle, the RILIS can be used for the ionization of almost all metallic elements and an important aspect of ongoing development is the extension of its range with the study of new ionization schemes. Ionization schemes for Sb, Sc, Dy and Y have been determined during recent off-line RILIS development. The atomic transitions between excited atomic states were studied by means of the resonance ionization spectroscopy at the RILIS setup. Based on the experimental results, optimal ionization schemes are defined.

2. The resonance ionization laser ion source

References [2, 3] give a thorough description of the ISOLDE RILIS. A master oscillator – power amplifier system of copper vapor lasers (CVL) provides two output laser beams in the form of 18 ns pulses at a repetition rate of 11 kHz, each with an average power of typically 30–40 W. The lasers rely on stimulated emission from two copper spectral lines, resulting in laser light comprising of green (511 nm) and yellow (578 nm) components. After separation of these components, four beams are available for the pumping of dye lasers and, where applicable, non-resonant ionization of atoms brought to a highly excited state by one or more previous resonant photon absorption steps. The RILIS set-up includes three dye lasers and therefore ionization schemes employing up to three resonant transitions can be used. The wavelength range of the dye lasers is 530–850 nm. Tuning is achieved by rotation of the diffraction grating in the laser resonator cavity and, depending on the diffraction grating used, the spectral width of the laser line is between 9 and 30 GHz. Frequency doubling and summation (tripling) is carried out using non-linear BBO (beta-barium borate) crystals to generate second or third harmonics of the fundamental beam, extending the wavelength range to include 214–415 nm. This enables high lying first excited atomic states to be accessed and is essential for elements with a high ionization potential. With this current work included, the RILIS has been used for resonance ionization of 24 of the elements. Schemes using one, two or three resonant transitions have been used. Most commonly, the ionization step is a transition to the continuum using an available CVL beam. Alternatively, the final step can be a resonant transition to an auto-ionizing state. A transition to an auto-ionizing state can have a high cross-section and therefore saturation is sometimes possible, improving both the ionization efficiency and the stability of the ion current.

The ionization takes place in a hot cavity connected to the target. Reaction products enter this cavity as an atomic vapour at a temperature of around 2,300 K. The role of the cavity is to contain the atoms for a certain time within a volume where they can be irradiated by the laser light and to confine the ions during their drift towards the extraction region. The ionization cavities are refractory metal (W or Nb) tubes with an inner diameter of 3 mm and a length of 30 mm. They are resistively heated to a temperature of about 2,300 K with a DC current of 200–350 A. After leaving the source, ions are accelerated to 60 kV, separated in a magnetic field and guided to the user by electrostatic ion-optical elements. For some elements nuclear effects such as hyperfine splitting and isotope shifts can reduce the RILIS efficiency. For the latter, the efficiency is maintained by small wavelength changes to allow for the shift of the resonance position along the isotope chain. Although detrimental to the overall RILIS efficiency, a large hyperfine splitting can enable preferential ionization of a particular isomer. This is particularly useful for nuclear spectroscopy experiments that benefit greatly from this spin selective ionization process. Isomer separation has been applied during on-line runs for Ag, Cu, Pb and Bi isotopes [3]. Furthermore, by operating in narrow band mode, where the laser bandwidth is reduced to close to that of the Doppler broadening of the atomic transitions ($\approx$ 1–3 GHz), the RILIS can be used as a precision spectroscopic tool for the study of these nuclear effects.

3. Ionization scheme development

Ionization scheme development at the RILIS has become a well established procedure comprising of four key steps:

3.1. LITERATURE SEARCH

After studying the various sources of atomic spectral line data, theoretical ionization schemes can be constructed. Some predictions about the relative efficiencies of the proposed schemes can be made if the data includes measurements of the line strengths or excited state lifetimes. Web based atomic spectral line databases [4] and a series of resonance ionization spectroscopy data sheets by E.B. Saloman [5, 6] have been particularly useful for this data collection procedure.

3.2. RESONANCE IONIZATION SPECTROSCOPY OF A STABLE BEAM

A small sample of the stable isotope is placed in an oven, which is attached to the target. Heating the oven releases the sample as an atomic vapour that effuses into the ionizer cavity. For the work described here, where three step schemes are investigated, resonance ionization spectroscopy (RIS) involves scanning the second step frequency across the range of possible transitions whilst the frequency of the first step transition is fixed. The ion current is monitored on a faraday cup detector.

3.3. SATURATION MEASUREMENT

For selected transitions, the dependence of the ion current *versus* the power of the laser beam was measured. The experimentally measured saturation power (which is deposited within the 3 mm diameter of the ion source cavity) gives an indication of minimal required power for RILIS operation without a loss of efficiency.

3.4. EFFICIENCY MEASUREMENT

The tabulated values for the RILIS efficiency represent a measure of the combined efficiency of ionization, extraction, and release from the mass separator. A small (few micrograms) mass tracer of the stable isotope is placed either in an empty target container or in the oven attached to the target. After tuning the lasers to the chosen ionization scheme, heating the oven begins the tracer evaporation and release into the ion source. The mass separator is tuned to the mass of the stable isotope and the transmitted ion current is monitored on a Faraday cup detector. After complete tracer evaporation, the total ion release, given by the integrated ion current is compared with the original sample size to give the RILIS efficiency. For elements that are readily surface ionized in the hot cavity, in this case, Sc and Y, the RILIS efficiency can be measured without the need to operate the lasers during the entire evaporation process. If the laser on/off ion current ratio is known for the various ion source and target heating settings used during the evaporation process, the RILIS efficiency can be estimated by multiplying the integral of the surface ion current by the appropriate laser on/off ratio.

4. New schemes for Sb, Sc, Dy and Y

The ionization potentials for Sc, Y and Sb (6.56, 6.22 and 8.61 eV, respectively) necessitate the use of three-step ionization schemes with the first transition at an energy corresponding to a laser wavelength in the UV region. This is outside the normal tuning range of the dye lasers and is achieved by frequency doubling or summation (tripling) of the fundamental laser light. For Dy, with an ionization potential of 5.94 eV, the first transition can be achieved with the fundamental frequency of the dye lasers.

The schemes tested all used non-resonant ionization steps excited by a CVL beam. This transition to the continuum has a low cross-section and, since saturation is not achieved, the ionization efficiency scales linearly with the available CVL power.

4.1. ANTIMONY

Antimony has a ground state configuration of $5s^25p^3\ {}^4S_{3/2}$ and four relatively strong transitions from the ground state to excited states lying between 43,000 and 49,000 cm^{-1} [7] are known. The transition to the $5p^26s\ {}^4P_{3/2}$ level at 45,945.34 cm^{-1} was chosen due to the abundance of known second step tran-

Table I. Summary of Sb spectroscopy

E_2, cm^{-1}	State II	ν_2, cm^{-1}	λ_2 (air), nm	Ion current (relative)	Laser power, mW[a]	
					1	2
60,765.29	6p $(^3P_0, 1/2)_{1/2}$	14,819.95	674.58	0.015	85	100
61,000.30	6p $(^3P_0, 3/2)_{3/2}$	15,054.96	664.05	0.12	110	300
62,462.41	4f $(^3P_0, 3)_{5/4}$	16,517.07	605.27	0.01	100	230
63,606.31	7p $(^3P_1, 1/2)_{1/2}$	17,660.97	566.06	0.3	–	240
63,790.95	**7p$(^3P_1, 3/2)_{5/7}$**	**17,845.61**	**560.21**	**1**	**60**	**300**
63,798.45	7p $(^3P_1, 1/2)_{3/2}$	17,853.11	559.97	0.88	–	–
63,900.53	7p $(^3P_1, 3/2)_{3/3}$	17,955.19	556.79	0.58	–	800
64,098.36	8p $(^3P_0, 1/2)_{1/2}$	18,153.02	550.72	0.14	–	600
64,209.43	7p $(^3P_1, 3/2)_{1/2}$	18,264.09	547.37	0.16	–	120
64,273.86	8p $(^3P_0, 3/2)_{3/2}$	18,328.52	545.45	0.08	–	90

The power of the CVL beam used for the non-resonant ionization step was 18 W (measured on the laser table).
[a]The values given in the tables are for the laser power measured on the laser table, transmission to the ion source is approximately 20% for the first step, 50% for the second step and 50% for the CVL beam (third step).

sitions from this level and, since this level shares the same multiplicity as the ground state, the transition was expected to be relatively strong. An energy of 45,945.34 cm^{-1} corresponds to a wavelength of 217.58 nm. The 652.77 nm fundamental beam was produced by pumping the Phenoxazone 9 dye with the yellow component of the CVL beam. The 3rd harmonic was generated using two BBO crystals to give the required UV wavelength.

Ten transitions between the $^4P_{3/2}$ excited state and known [8] higher atomic levels (60,765–64,274 cm−1) were observed in the wavelength range of 545–675 nm (Table I). The green component of the CVL beam was transferred to the ion source for the ionization step. The measurement of the ion current determined the optimal second step transition of 17,845.61 cm−1 to the $5p^27p$ $(^3P_1, 3/2)_{5/2}$ level at 63,790.95 cm−1. The laser power generated at this wavelength was between 200 and 300 mW, at least two times greater than the saturation power of approximately 100 mW. With an ionization potential of 8.61 eV, surface ionized antimony is not seen from the hot cavity and so the lasers remained on during the mass marker evaporation. From the evaporation of the Sb mass marker the RILIS efficiency was measured to be 2.7%.

4.2. SCANDIUM

Scandium has a $3d^4s^2$ $^2D_{3/2}$ ground state and a low lying excited state exists at 168.34 cm^{-1}. In the ionizer cavity at a temperature of 2,300 K the latter is 57.4% populated and so the spectroscopy study considered first step transitions from this level only. Due to the abundance of second step transitions [9, 10] and for

Table II. Scandium three-step schemes with E_1 = 307,066.66 cm^{-}1; State I = 3d4s(^{3}D)4p $^2P_{3/2}$

E_2, cm^{-1}	State II	ν_2, cm^{-1}	λ_2 (air), nm	Laser power, mW			Las/surf ion ratio
				1	2	3	
44,060.44	new	13,353.8	748.65	–	170	1,000	5.5
44,496.16	new	13,789.5	724.99	–	240	–	41.7
44,594.97	**3d4p^2 $^2P_{3/2}$**	**13,888.3**	**719.83**	**200**	**200**	**1,000**	**400**
44,690.65	3d4p^2 $^2P_{3/2}$	13,984.0	714.91	–	200	–	352.9
45,514.98	3d4s4d $^2S_{1/2}$	14,808.3	675.11	70	60	1,000	216.7
45,672.16	new	14,965.5	668.02	–	170	–	120.7
45,764.56	new	15,057.9	663.92	–	–	–	36
45,824.26	new	15,117.6	661.30	–	90	900	28.3
45,866.86	new	15,160.2	659.44	–	100	900	41.7
46,027.56	new	15,320.9	652.52	–	140	–	–
46,563.76	new	15,857.1	630.46	–	130	–	–
46,652.96	new	15,946.3	626.93	–	110	–	51.9
46,825.26	new	16,118.6	620.23	–	70	1,000	42.3
46,924.76	new	16,218.1	616.42	–	70	1,000	92
47,425.46	$^2D_{5/2}$	16,718.8	597.96	–	120	900	125
47,588.36	new	16,881.7	592.19	–	150	950	161.8
47,626.46	new	16,919.8	590.86	90	120	1,000	448.3
47,723.96	new	17,017.3	587.47	–	170	900	84.4
47,761.86	new	17,055.2	586.17	–	170	800	20.3
47,783.86	new	17,077.2	585.41	–	150	800	243.3
47,794.36	new	17,087.7	585.05	–	150	800	106.5
47,998.76	new	17,292.1	578.14	–	200	900	77.4
48,065.46	new	17,358.8	575.92	–	320	1,000	121.6
48,065.76	new	17,359.1	575.91	–	150	800	173.3
48,065.76	new	17,359.1	575.91	100	130	800	250
48,107.76	new	17,401.1	574.52	–	320	1,000	114.3
48,229.56	new	17,522.9	570.52	–	100	1,000	39.1
48,229.66	new	17,523	570.52	70	300	1,000	59.1
48,490.16	new	17,783.5	562.16	100	200	1,000	325.6
48,920.6	$^4P_{5/2}$	18,213.94	548.88	–	190	1,000	5
48,975.46	new	18,268.8	547.23	–	160	1,000	7.5
49,069.96	new	18,363.3	544.41	–	120	1,000	13.5
49,146.46	new	18,439.8	542.15	–	100	1,000	12.4

convenience in terms of wavelength tuning, we chose a first step transition to the 3d4s(^{3}D)4p $^2P_{3/2}$ state at 30,706.66 cm^{-1}. The UV (327.36 nm) beam required for this transition was generated by frequency doubling of the fundamental beam obtained by pumping the Phenoxazone 9 dye with the yellow component of the CVL beam. The RIS work involved scanning the second step frequency from this first step transition, covering the energy level range 44,030–49,210 cm^{-1}. Five laser dyes (Rhodamine 110, Pyrromethene 597, DCM, Phenoxazone 9, and Rhodamine 700) were used to generate the wavelengths in the spectral range of

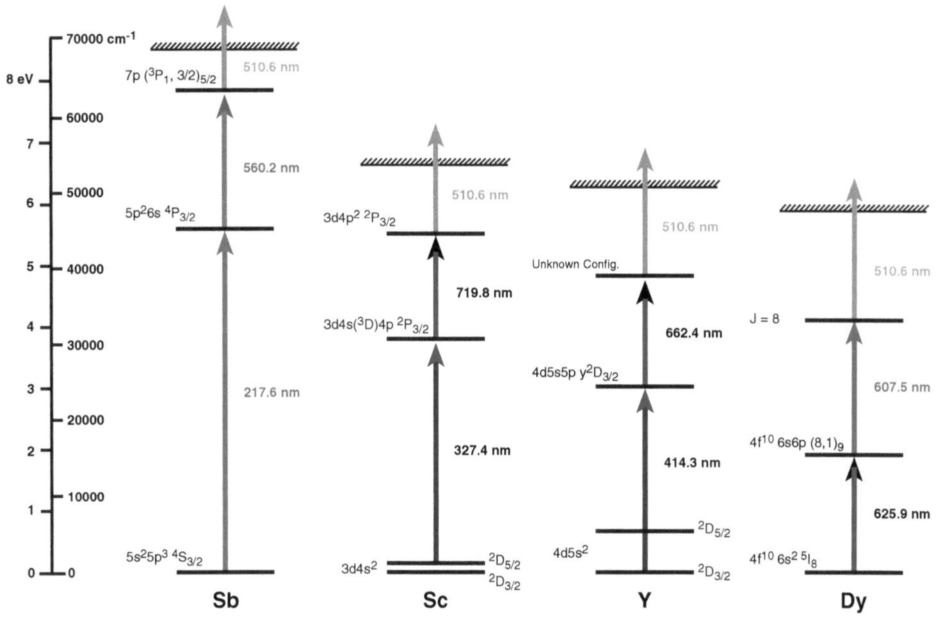

Figure 1. Optimal RILIS ionization schemes.

540–750 nm. Within this region 26 transitions to new excited states were observed and 27 transitions to documented levels [10] from the $^2P_{3/2}$ state were also seen. Table II gives a summary of the efficient schemes measured and also the newly observed atomic transitions. Highlighted in bold is the most efficient scheme, as judged by the laser on/off ion ratio. The power available for the first and second step transitions for this scheme was greatly in excess of the measured saturation powers (5 and 20 mW, respectively). The optimal scheme (Figure 1), uses a second step transition to the $3d4p^2\ ^2P_{3/2}$ level at 44,594.57 cm^{-1} and gives a laser to surface ion ratio of 400 at typical target and line temperatures.

For the efficiency measurement the lasers were not used and the sample was surface ionized in the hot cavity. The complete sample evaporation required a large increase in the target temperature (from 500 to 830 A) and almost 0.1% of the 1,650 nAh sample was collected. Due to secondary heating of the ionizer cavity (by conduction and radiation from the target), the laser/surface ion ratio of 400 (Table II), measured at a target current of 570 A, is not a reliable factor for use in calculating the RILIS efficiency. Although the dependence of the laser/surface ion ratio on the target temperature was not measured for scandium, this effect was measured as part of the off-line spectroscopy work for dysprosium. A 50% reduction in the laser/surface ion ratio was observed with a 150 A increase in the target temperature. Based on this exponential relation observed for Dy, an estimate of the target temperature dependent laser/surface ion ratio for Sc can be

Table IIIA. Yttrium three-step schemes

E_2, cm^{-1}	State II	ν_2, cm^{-1}	λ_2 (air), nm	Laser power, mW		Las/surf ion ratio
				1	2	
39,001.4	new	14,870.2	672.30	110	250	8.7
39,087.2	new	14,956.0	668.44	110	400	5.7
39,209.3	new	15,078.1	663.03	110	500	12.9
39,224.4	new	15,093.2	662.37	110	600	88.4
39,313.2	new	15,182.0	658.50	110	740	51.8
39,446.3	4d5s5d $e^4F_{3/2}$	15,315	652.77	110	750	22.1
39,553.0	new	15,421.8	648.25	110	900	22.3
39,565.1	4d5s5d $e^4F_{5/2}$	15,433.8	647.75	110	870	24.3
39,686.0	new	15,554.8	642.71	110	920	30.9
40,287.6	new	16,156.4	618.78	110	720	9.4
40,307.7	new	16,176.5	544.41	110	700	11.5
41,423.1	new	17,291.9	578.15	110	1200	25.8
41,660.3	new	17,529.1	570.32	110	1600	32.9
41,660.3	new	17,529.1	570.32	110	1400	31.2
41,669.5	new	17,538.3	570.02	110	1500	16.1
41,853.3	new	17,722.1	564.11	110	1400	31.6
41,879.9	new	17,748.7	563.27	110	900	7
41,992.8	new	17,861.6	559.70	110	700	2.3

$E_0 = 0$ cm^{-1}; $E_1 = 24{,}131.2$ cm^{-1}; State I = 4d5s5p $y^2D_{3/2}$.

made. The RILIS efficiency, calculated using this variable laser/surface ion ratio is estimated at 15%.

4.3. YTTRIUM

The structure of atomic levels of yttrium is similar to that of scandium as both belong to the same group in the periodic table of elements. The ground state of Y $4d5s^2$ ^2D is split on two levels with $J = 3/2$ and 5/2, where the 4d $5s^2$ $^2D_{5/2}$ excited state, lies at an energy of 530.36 cm^{-1} and is 52% populated. The available data on the atomic transition probabilities of yttrium [11] presents a choice of first step transitions for the resonance ionization. Schemes using relatively strong transitions from each of these levels to five first excited states with the electronic configuration 4d5s5p between 24,100 and 25,000 cm^{-1} (407–415 nm) were tested during the RIS study. By scanning the second step within the wavelength range of 565–617 nm, 45 highly excited states were observed in the energy range of 39,000–43,400 cm^{-1}. Among these highly excited states only eight levels were known from the atomic energy level tables [12]. Laser ion/surface ion ratios were measured for 72 schemes, details of which are given in Tables III (A–E). The saturation power for the chosen first step transition was approximately 10 mW, an order of magnitude lower than the power available during the measurements.

Table IIIB. Yttrium three-step schemes

E_2, cm^{-1}	State II	ν_2, cm^{-1}	λ_2 (air), nm	Laser power, mW		Las/surf ion ratio
				1	2	
39,313.2	new	14,832.6	674.01	170	200	22
39,322.1	new	14,841.5	673.60	170	–	0.4
39,446.3	4d5s5d $e^4F_{3/2}$	14,965.7	668.01	170	400	16.5
39,552.8	new	15,072.2	663.29	170	440	15.8
39,565.1	4d5s5d $e^4F_{5/2}$	15,084.2	662.76	170	400	13.5
39,686.0	new	15,205.4	657.48	170	640	17.6
40,287.6	new	15,807.0	632.46	170	840	13.5
40,307.7	new	15,827.1	631.65	170	940	14.1
40,455.1	4d5s5d $f^4P_{3/2}$	15,974.3	625.83	170	900	0.3
40,517.1	4d5s5d $f^4P_{5/2}$	16,036.35	623.41	170	800	16
41,660.6	new	17,180.0	581.91	140	780	40.1
41,669.7	new	17,189.1	581.60	140	–	31.9
41,853.3	new	17,372.7	575.46	140	1,200	27
41,880.3	new	17,399.7	574.56	140	1,300	16.1
42,048.9	new	17,568.3	569.05	140	1,500	34.7
42,098.5	new	17,617.9	567.45	140	1,400	41.9
42,106.6	new	17,626.0	567.19	140	1,400	16.9
42,135.0	new	17,654.4	566.27	140	–	2.5
42,171.0	new	17,690.4	565.12	140	1,300	11.9
42,220.9	new	17,740.3	563.53	140	1,200	1.5
42,253.3	new	17,772.7	562.50	140	1,100	39
42,301.3	new	17,820.7	560.99	140	800	1.1

E_0 = 0 cm^{-1}; E_1 = 24,480.60 cm^{-1}; State I = 4d5s5p $y^2P_{3/2}$.

4.4. DYSPROSIUM

The $4f^{10}\,6s^2\,{}^5I_8$ ground state of Dy is 93% populated in the ionizer cavity and 12 accessible excited states exist between 13,500 and 18,500 cm^{-1}, corres-

Table IIIC. Yttrium three-step schemes

E_2, cm^{-1}	State II	ν_2, cm^{-1}	λ_2 (air), nm	Laser power, mW		Las/surf ion ratio
				1	2	
39,313.2	new	14,794.4	675.75	160	100	8.9
39,446.3	4d5s5d $e^4F_{3/2}$	14,927.5	669.72	160	300	11.1
39,553	new	15,034.2	664.97	160	520	17.1
39,565.1	4d5s5d $e^4F_{5/2}$	15,046.3	664.43	160	520	22.1
39,686	new	15,167.2	659.14	160	650	33.1
39,757.8	4d5s5d $e^4F_{7/2}$	15,239.1	656.03	160	800	9.9
40,517.1	4d5s5d $f^4P_{5/2}$	15,998.3	624.89	160	700	12.9

E_0 = 0 cm^{-1}; E_1 = 24,518.80 cm^{-1}; State I = 4d5s5p $y^2F_{5/2}$.

Table IIID. Yttrium three-step schemes

E_2, cm^{-1}	State II	ν_2, cm^{-1}	λ_2 (air), nm	Laser power, mW		Las/surf ion ratio
				1	2	
39,565.1	4d5s5d $e^4F_{5/3}$	14,818.5	674.65	120	150	30.8
39,686.0	new	14,939.4	669.19	120	420	27.6
39,757.8	4d5s5d $e^4F_{7/2}$	15,011.2	665.99	120	600	8.6
40,455.1	4d5s5d $f^4P_{3/2}$	15,708.5	636.42	120	–	1.9
40,517.1	4d5s5d $f^4P_{5/2}$	15,770.5	633.92	120	1200	20.7
41,722.4	new	16,975.8	588.91	140	380	7.9
41,853.3	new	17,106.7	584.40	140	700	10.1
42,048.8	new	17,302.2	577.80	140	1,200	16.4
42,098.6	new	17,352.0	576.14	140	–	24.5
42,106.6	new	17,360.0	575.88	140	1,300	25.5
42,220.8	new	17,474.2	572.11	140	–	3.1
42,253.6	new	17,507.0	571.04	140	1,500	23.7
42,391.3	new	17,644.7	566.59	140	1,300	4.1
42,487.9	new	17,741.3	563.50	140	1,200	13.5

E_0 = 530.36 cm^{-1}; E_1 = 24,746.60 cm^{-1}; State I = 4d5s5p $y^2D_{5/2}$.

ponding to a photon transition in the visible region, within the tuning range of the fundamental beam from the dye laser. Using the Phenoaxazone 9 dye pumped with the yellow component of the CVL beam, we measured schemes using three of these levels close to 15,500 cm^{-1} and for transitions from these levels, 88 second excited states are documented. Using the green component of the CVL beam for non-resonant ionization and by scanning the second step frequency over the spectral range of the Phenoaxazone 9 and DCM dyes (607–

Table IIIE. Yttrium three-step schemes

E_2, cm^{-1}	State II	ν_2, cm^{-1}	λ_2 (air), nm	Laser power, mW		Las/surf ion ratio
				1	2	
39,686.0	new	14,786.5	676.11	120	70	0.1
39,757.8	4d5s5d $e^4F_{7/2}$	14,858.3	672.84	120	260	0.9
39,963.7	4d5s5d $e^4F_{9/2}$	15,064.2	663.64	120	400	17.4
40,189.2	new	15,289.7	653.85	120	780	0.1
40,429.9	new	15,530.4	643.72	120	1,100	0.2
40,517.1	4d5s5d $f^4P_{5/2}$	15,617.6	640.13	120	1,200	2
42,098.6	new	17,199.1	581.26	110	810	29.5
42,253.3	new	17,353.8	576.08	110	–	1.6
42,390.9	new	17,491.4	571.55	110	–	5.3
42,487.6	new	17,588.1	568.41	110	–	28.1

E_0 = 530.36 cm^{-1}; E_1 = 24,899.50 cm^{-1}; State I = 4d5s5p $y^2F_{7/2}$.

Table IVA. Dysprosium three-step schemes

E_2, cm^{-1}	State II	ν_2, cm^{-1}	λ_2 (air), nm	Laser power, mW		Las/surf ion ratio
				1	2	
30,979.48	$4f^{10}$ 6s7s $(8, 1)_8$	15,784.70	633.35	200	300	0.51
30,988.16	$J = 6$	15,793.42	633.00	200	300	1.21
31,180.01	$J = 6$	15,985.18	625.41	200	600	0.27
31,362.62	$J = 7$	16,167.79	618.34	200	1,000	7.82
31,423.04	$J = 7$	16,228.21	616.04	200	950	0.83
31,469.00	$J = 8$	16,274.17	614.30	200	1,100	8.8
31,471.44	new	16,276.61	614.21	200	1,000	6.4
31,509.07	$4f^{10}$ 6s7s $(8, 1)_7$	16,314.24	612.79	200	950	2.98
31,544.68	new	16,349.85	611.46	900	1,000	16.45
31,619.03	new	16,424.20	608.69	200	800	8.8
31,654.20	new	16,459.37	607.39	200	650	0.29
31,674.01	$J = 7$	16,479.18	606.66	200	550	5.86
31,684.36	new	16,489.53	606.28	200	500	9.78

$E_0 = 0$ cm^{-1}; $E_1 = 15{,}194.83$ cm^{-1}; State I = $4f^9$ $5d6s^2$ $^5H_7$.

680 nm), schemes using 19 of the known second excited states, and also seven new levels, were measured. A total of 35 ionization schemes were measured, the details of which are given in Tables IV (A–C). Saturation of the three first step and four of the second step transitions corresponding to the strongest schemes was confirmed. The optimal scheme (shown in Figure 1) gave a laser/surface ion ratio of close to 60 with the target and line heating at 500 and 310 A, respectively.

Dy has seven stable isotopes from mass 156 to 164 and readily forms an oxide (DyO) or a fluoride (DyF). The corresponding isotope and molecular masses

Table IVB. Dysprosium three-step schemes

E_2, cm^{-1}	State II	ν_2, cm^{-1}	λ_2 (air), nm	Laser power, mW		Las/surf ion ratio
				1	2	
31,362.62	$J = 7$	15,795.24	632.93	1,100	300	4.19
31,423.04	$J = 7$	15,855.66	630.52	1,100	450	6.16
31,469.00	$J = 8$	15,901.62	628.69	1,100	520	3.46
31,509.12	$4f^{10}$ 6s7s $(8, 1)_7$	15,941.74	627.11	1,100	600	7.93
31,654.34	new	16,086.96	621.45	1,100	900	0.22
31,674.08	$J = 7$	16,106.70	620.69	1,100	950	0.91
31,694.88	new	16,127.50	619.89	1,100	1,000	0.47
31,775.65	$J = 9$	16,208.27	616.80	1,100	1,000	26.27
31,820.28	$J = 8$	16,252.90	615.10	1,100	1,100	2.11
32,036.51	$J = 7$	16,469.13	607.03	1,100	600	13.42

$E_0 = 0$ cm^{-1}; $E_1 = 15{,}567.38$ cm^{-1}; State I = $4f^{10}$ 6s6p $(8,0)_8$.

Table IVC. Dysprosium three-step schemes

E_2, cm^{-1}	State II	ν_2, cm^{-1}	λ_2 (air), nm	Laser power, mW		Las/surf ion ratio
				1	2	
30,739.79	$J = 8$	14,767.44	676.98	800	40	2.23
30,979.53	$4f^{10}$ 6s7s $(8, 1)_8$	15,007.18	666.16	800	150	2.32
31,061.18	$J = 8$	15,088.83	662.56	750	100	0.78
31,233.57	$J = 8$	15,261.22	655.07	650	800	51.43
31,287.04	$J = 9$	15,314.69	652.79	750	160	0.66
31,469.00	$J = 8$	15,496.65	645.12	750	140	0.64
31,489.64	$J = 10$	15,517.29	644.26	600	800	2.99
31,775.37	new	15,803.02	632.62	650	800	4.96
31,820.12	$J = 8$	15,847.77	630.83	–	800	3.21
31,838.26	$J = 10$	15,865.91	630.11	–	680	21.4
32,292.38	$J = 8$	16,320.03	612.57	550	1,000	4.37
32,428.66	$J = 8$	16,456.31	607.50	800	550	57.25

$E_0 = 0$ cm^{-1}; $E_1 = 15{,}972.35$ cm^{-1}; State I = $4f^{10}$ 6s6p $(8,1)_9$.

were selected in turn and the relative yields of each were measured. With the exception of ^{158}Dy, the relative yields of each isotope closely matched their relative natural abundances, indicating that the isotope shift of the atomic transition is not significant across the stable isotope chain.

During the efficiency measurement a 300 μg (12,620 nAh ^{162}Dy) sample was evaporated at a line heating of 300 A with the lasers turned off and with the target temperature increasing from 500 to 820 A. As discussed in Section 4.2, such an increase in target temperature results in significant secondary heating of the line and, in turn, an enhancement of the surface ionization efficiency. The expected charge accumulation by laser ionization was estimated by multiplication of the surface ion current by a varying laser/surface ion ratio appropriate to the target and line temperature at each stage of the evaporation. 0.4% of the sample was collected after surface ionization and the RILIS efficiency was evaluated as 20%.

5. Conclusion

As the scope of the RILIS has expanded to include more of the elements, the use of RILIS produced radioactive ion beams has greatly increased and now accounts for over half of the on line experimental shifts at ISOLDE. During 2003 the RILIS was in operation for a total of 1,810 h, with only a small proportion of this time set aside for development work (240 h). The technique of laser resonance ionization can in principle be applied to almost all metallic elements however, at ISOLDE where reaction products are stopped in a thick target matrix, the subsequent diffusion and effusion processes greatly inhibit the release of refractory elements. For this reason, RILIS scheme development at ISOLDE is limited to the more volatile metals according to the requests of the users and possible physics interests. It is anticipated that further work of this kind at ISOLDE will include the investigation of new ionization schemes for Po, Tl, Ge, Hg and Au.

Acknowledgements

For the tracer mass markers, the sample preparation and installation was the work of Richard Catherall and Bernard Crepieux of the ISOLDE Collaboration.

References

1. Mishin V. I., Fedoseyev V. N., Kluge H.-J., Letokhov V. S., Ravn, H. L. Scheerer F., Shirakabe Y., Sundell S., Tengblad O. and the ISOLDE Collaboration, *Nucl. Instrum. Methods Phys. Res. B* **73** (1993), 550.
2. Fedoseyev V. N., Huber G., Kster U., Lettry J., Mishin V. I., Ravn H. L., Sebastian V. and the ISOLDE Collaboration, *Hyperfine Interact.* **127** (2000), 409.
3. Fedosseev V. N., Fedorov D. V., Horn R., Huber G., Kster U., Lassen J., Mishin V. I., Seliverstov M. D., Weissman L., Wendt K. and the ISOLDE Collaboration, *Nucl. Instrum. Methods Phys. Res. B* **204** (2003), 353.
4. http://physics.nist.gov/cgi-bin/AtData/main_asd
5. Saloman E.B., *Spectrochim Acta Part B* **47** (1992), 517.
6. Saloman E. B., *Spectrochimica Acta Part B* **49** (1994), 251.
7. Zaidi A. A., Makdisi Y. and Bhatia K. S., *J. Phys. B: At. Mol. Phys.* **17** (1984), 355.
8. Hassini F., Ben Ahmed Z., Robaux O., Vergs J. and Wyart J.-F., *J. Opt. Soc. Am. B* **5** (1988), 2060.
9. Sugar J. and Corliss C., *J. Phys. Chem. Ref. Data* **9** (1980), 473.
10. Kaufman V. and Sugar J., *J. Phys.Chem. Ref. Data* **17** (1988), 1679.
11. Reshetnikova O. F. and Skorohod E. P., *Opt. Spectrosc.* **87** (1999), 911.
12. Moore C.E., *Atomic energy levels*, NSRDS-NBS 35 (U.S. Government Printing Office, Washington, District of Columbia, 1971), Vol. II.

Hyperfine Interactions (2005) 162:29–38
DOI 10.1007/s10751-005-9207-z

Development Towards a Laser Ion Source Trap for the Production of Exotic Species

K. WIES[1,*], C. GEPPERT[1], K. BLAUM[1], K. BRÜCK[1], H.-J. KLUGE[2], S. SCHWARZ[3] and K. WENDT[1]
[1]*Institut für Physik, Johannes Gutenberg-Universität, 55099 Mainz, Germany*
e-mail: Katja.Wies@uni-mainz.de
[2]*GSI Darmstadt, Planckstraße 1, 64291 Darmstadt, Germany*
[3]*NSCL, Michigan State University, East Lansing, MI 48824-1321, USA*

Abstract. A new type of resonance ionization laser ion source (RILIS) is presently being developed and tested at the off-line mass separator at Mainz University for future use at on-line exotic rare isotopes production facilities. For highest isobaric selectivity, this RILIS approach decouples the evaporation and ionization process. A further advantage is the generation of full temporal control of the resulting high quality ion beam. These facts are realized by a combination of atomizer – ion repeller – ion cooler and trap, which is operated together with a state-of-the-art, all solid state laser system. The principle and performance of this laser ion source trap (LIST) system are discussed applying simulation studies for the repeller-trap combination and first measurements for characterization.

Key Words: ion trap, laser ion source, radioactive ion beams, radio-frequency quadrupole, resonance ionization, SIMION, titanium:sapphire laser.

1. Introduction

In the last 10 years laser resonance ionization has developed into a mature technique. In particular it plays a key role for the production of radioactive nuclides at on-line isotope separator facilities (ISOL). At the on-line mass separator facility ISOLDE / CERN, Geneva, laser ionization ion sources (RILIS) nowadays are the most commonly used source type [1], while other facilities, i.e., ISAC, TRIUMF, Vancouver or JYFL, University of Jyväskylä, Finland, are currently setting up corresponding laser systems and developing source geometries.

Due to the applied multi-step resonant excitation and ionization process with laser light, RILIS has the potential of providing very high elemental selectivity and could produce isobarically clean ion beams. Unfortunately, due to the required high temperatures around ~2,000 K at the on-line atomizer, some

* Author for correspondence.

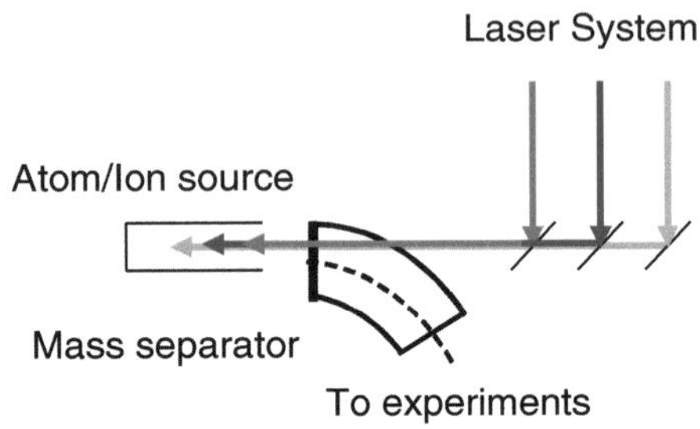

Figure 1. Sketch of a laser ion source. Atoms from an atomic beam source are resonantly excited and ionized by laser light. The produced photo-ions are mass separated and finally detected or sent to experiments.

surface ionization of unwanted atomic species occurs and limits the isobaric selectivity. In the setup of a laser ion source trap (LIST), as first described in [2], this limitation is overcome by decoupling the evaporation and laser ionization through one or more repeller electrodes. By applying corresponding potentials, unwanted charged species like surface ions as well as electrons from the atomizer are rejected. Selective laser ionization of the evaporated neutral atoms is carried out behind the repeller and inside a radio-frequency quadrupole (RFQ) ion trap. The trap system is filled with buffer-gas and segmented, so that it can be used for accumulation, cooling and storage of the laser ions. By pulsed extraction of the stored ion cloud, the system provides temporal control of ion bunches extracted towards the isotope separator. In addition to highest elemental purity and excellent beam quality, high beam brilliance during the pulse significantly increases signal to background ratios for numerous experimental arrangements (i.e., collinear laser spectroscopy) and perfectly adapts to recollection of the ions for subsequent ion trap experiments. Thus it is expected that a wide range of the experiments installed at ISOL systems will benefit from the advantages of this novel RILIS concept.

2. Experimental setup

A schematic setup of a laser ion source installation at an isotope separator is shown in Figure 1. It consists of three main components, the atom/ion source with optional trap arrangement, the isotope separator and the laser system, which shall be described in opposite order below:

2.1. LASER SYSTEM

Requirements for highest ionization efficiency and elemental selectivity, which are almost independent of the element of choice, determine the specifications of an ideal laser system for the purpose of an on-line ion source. In addition reliable long-term operation, negligible maintenance and highest versatility should be considered. To best meet all these demands, a high repetition rate all-solid-state nanosecond laser system has been set up. The second harmonic of an arc-lamp-driven Nd:YAG laser (CLARK, ORC-1000) with an average laser power of up to 40 W serves for pumping three tunable titanium:sapphire (Ti:Sa) lasers. A Z-shaped resonator with a combination of a Lyot-filter and an etalon is used for generation of narrow bandwidth laser radiation (~5 GHz); frequency selection is computer-controlled via variation of the etalon tilt. For temporal synchronization of the three Ti:Sa lasers, as needed for efficient multi-step excitation, a remote-controlled Q-switch is installed in each individual resonator. At a typical repetition rate of 10 kHz, each one of these lasers provides about 2 W of average power with 30–50 ns pulse length [3]. Via the spectral range of four different mirror sets, the wavelength regime from 725 to 925 nm in the infrared spectral region is accessible. Frequency doubling and tripling into the blue and ultraviolet spectral regions is achieved by second and third harmonic generation within non-linear optical crystals. In this way, fully resonant three step resonance ionization schemes for more than 80 elements are accessible. By either using autoionization or high-lying atomic Rydberg states together with subsequent field ionization in the electric fields of the ion trap or the isotope separator, usually most of the steps of an excitation ladder can be saturated and highest ionization efficiency is realized. A more detailed description of the basics and layout of this laser setup will be given elsewhere [4].

2.2. ISOTOPE SEPARATOR

The RISIKO isotope separator (similar to ISOLDE II [5]) at Mainz University is described in [6] and references therein. It is operated with a directly heated oven for atomization and subsequent laser ionization or alternatively for direct surface ionization of the sample material, depending on oven material and temperature. Ion beams are accelerated by a two-stage extraction ion optics to a beam energy of 30–60 keV. Mass separation is carried out in a double focusing 60° dipole magnet (Danfysik) with 1 m curvature and focal length. A straight-through window in the magnet yoke and vacuum chamber provides an entrance for coupling of the laser beams into the RILIS. The magnet focal plane is equipped with a slit system, after which the mass separated ions are detected as current in a Faraday cup or counted in a channeltron detector.

2.3. RADIO-FREQUENCY QUADRUPOLE TRAP

The central part of the atomizer – ion repeller – ion cooler and trap combination is a segmented and gas-filled radio-frequency quadrupole ion trap. These devices are nowadays routinely used for cooling, accumulation and bunching of ion beams [7, 8]. A cross-section through the oven-repeller-trap combination is

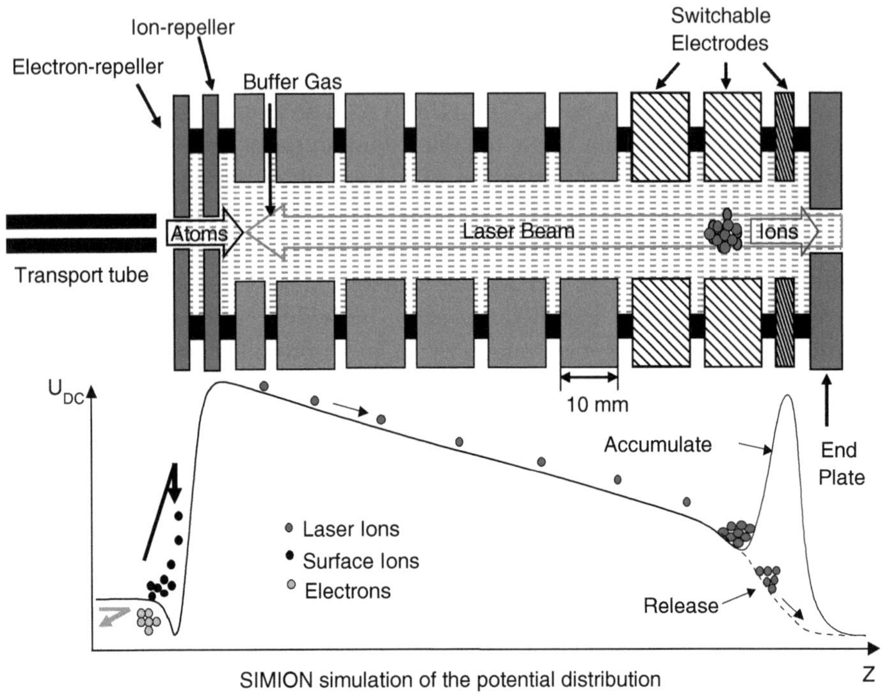

Figure 2. Working principle and potential distribution of the repeller-trap system. Atomization and ionization processes are decoupled by two-repeller electrodes. Photo ions are created inside the trap by laser light. The produced ions are trapped and cooled by He buffer gas. After a cooling time the ions are ejected as an ion bunch.

shown in Figure 2 together with a typical potential distribution applied for trapping. The atoms are delivered through the transport tube from the location of evaporation, acting as an atomic beam source. Two repeller electrodes serve for suppression of electrons and unwanted surface ions from the source. Each of them has an aperture diameter of 6 mm and a thickness of 2 mm. They are separated by 1 mm ceramic insulator disks. The RFQ rods have a diameter of 18 mm, a total length of 96 mm and consist of nine segments. The spacing between opposite rods is 12 mm. The length of the individual segments varies from 2 to 10 mm. The first and the last segments are shorter to allow for a sharp increase of the potential. Ceramic insulators keep neighboring segments 2 mm apart. A final electrode with an inner diameter of 6 mm and a thickness of 5 mm shields the internal of the trap against the acceleration field of the isotope separator. Additional hollow rods installed parallel to the RFQ rods provide buffer gas supply and stabilization, while ceramic insulator rods seal the trap section. The RF-frequency used is set to $\nu_{RF} = 1$ MHz with an amplitude of up to $U_{RF} = \pm 200$ V. DC-voltages in the order of up to 100 V are supplied for the longitudinal storage field and the repelling electrodes.

3. Simulation studies and results

3.1. SIMULATION STUDIES

The current design of the LIST has been determined by carefully fine tuning the trap parameters within simulation studies. A combination of the LAPLACE-solver SIMION 3D Version 7.0 [9] and the Monte-Carlo program LISBUN, written by one of the authors [10], was used. While the potential distribution of different geometries were analyzed and optimized using the SIMION code, the LISBUN code calculated the motion of ions in the presence of buffer-gas by microscopically simulating collisions with buffer gas atoms using realistic interaction potentials. Cesium-ions were considered as realistic species for mid-mass range. Simulations were done in two steps: first the initial distribution of the atoms for laser ionization was generated and afterwards the behaviour of the resulting photo-ions in the trap was studied with SIMION and LISBUN.

Starting conditions considered that atoms are produced in an atom source, a tube of 30 mm length and 3 mm diameter. It was estimated that the atoms leave this tube with an average velocity of about v = 500 m/s. Thus the mean atom flight path between two laser pulses at 10 kHz repetition rate is about 50 mm, half the length of the trap. Realistic beam divergences were obtained from transversal laser spectroscopic absorption experiments. The emitted atom ensemble is well described by the sum of two Maxwell velocity distributions, one with a small opening angle of about ±4° and one with a broad opening angle of about ±20°. Different temperatures and operation parameters of the atomizer primarily change the ratio of these two contributions but not the general geometry. In the simulations these two components were realized with one part of the atoms starting at the rear end inside the atom source and the other one starting at the front end of the oven tube. For both components a flat angular distribution was assumed for simplifying the numerical simulation calculations. Figure 3 shows the geometry indicating both distributions of atoms schematically and giving the potentials applied. Having transmitted the ion repeller, most of the atoms are ionized inside the RFQ trap at a random position along the first 50 mm and within the laser volume of 6 mm diameter. The axial distribution and the phase space volume of the created photo ions are shown in Figure 4. As a typical value 66% of the emitted atomic beam is assumed in the broad angle distribution and 34% in the narrow angle component. Under these conditions simulations give an efficiency of ~30% for ionization inside the trap, if saturation of all excitation and ionization steps is assumed. All atoms ionized inside the trap are trapped and cooled down by buffer gas collisions. The phase space diagram of Figure 5 shows the motion of an ion cloud in the axial direction of the RFQ trap at a helium (He) pressure of 1×10^{-2} mbar. Shortly after ionization the ions fill the whole trap (Figure 5a). They move towards the exit until they are reflected by the potential wall at z = 125 mm. As they lose energy in collisions with buffer gas atoms they do not reach the initial starting point on their way back and form

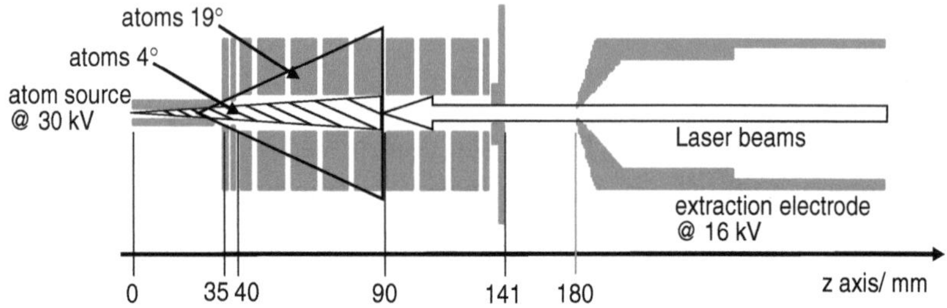

Figure 3. Calculated SIMION geometry with typical voltages: two atoms groups, one with starting angle of ±4° and one with ±19°, as considered as realistic starting conditions. In the simulations typical voltages of the Mainz off-line test setup have been used.

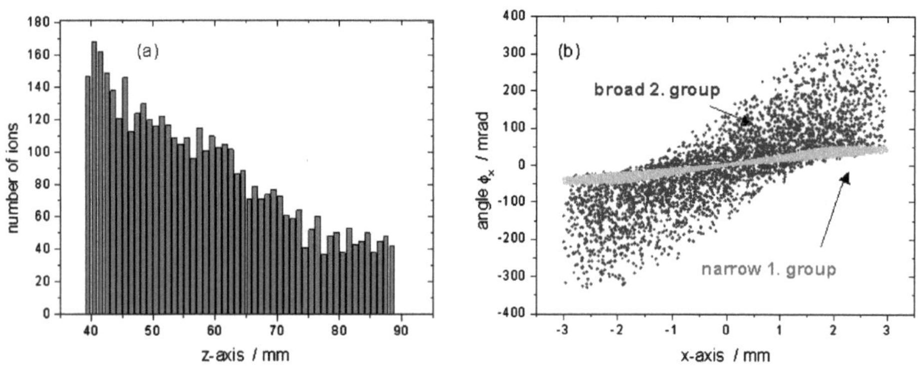

Figure 4. Simulation of photo ion creation: (a) axial photo ion distribution within the laser volume and (b) phase space of the two groups of ions.

a closed ion cloud (Figure 5b). After a cooling time of several ms, the size of the phase space occupied has reduced considerably and room temperature is reached with a well defined ion cloud of 0.5 mm diameter and 1.5 mm length (Figure 5c). The estimated efficiency for cooling and bunching is 98%, neglecting any space charge effects so far (see below). For a cooling time of 1 ms and a laser repetition rate of 10 kHz, ions from up to 10 laser pulses are accumulated in the trap. Figure 6 shows the resulting transversal emittances for an ion bunch created either by accumulating only one or alternatively 10 consecutive laser pulses within a cooling time of 1 ms with parameters set to 16 kV extraction potential and a He pressure of 1×10^{-2} mbar. The calculations give an 80% emittance for accumulated ions from 10 laser pulses of 5.7 π mm mrad, while single pulse accumulation reaches 2.2 π mm mrad. Both specifications are significantly better than the value of ~10 π mm mrad calculated for the laser ion source without trap. Temporal duration of the well-cooled ion beam pulse of ~50 ns is estimated for a primary acceleration by 16 kV. Setting the extraction efficiency to 100%, an overall efficiency for ionization, cooling, bunching and release of up to 30% is obtained. One limiting factor at low repetition rates is arising from the maximum

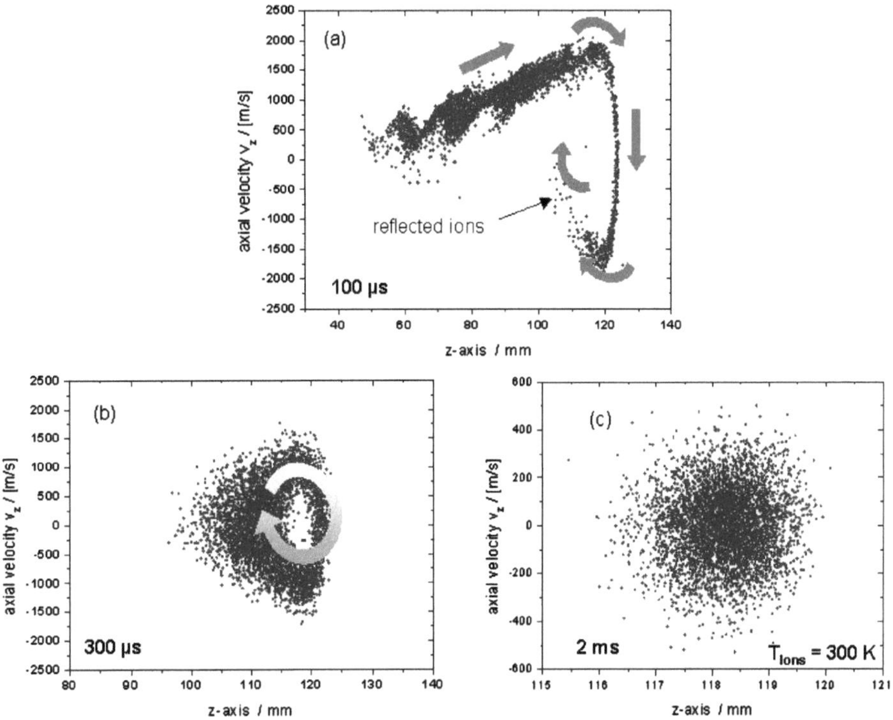

Figure 5. Temporal development of photo-ions inside the RFQ-trap.

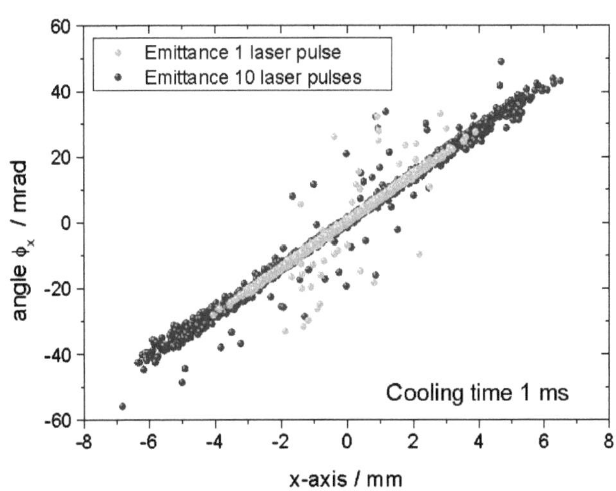

Figure 6. Transversal emittance of ions created by one and by ten laser pulses after a cooling time of 1 ms.

loading capacity of the trap, which is determined by space charge and has been calculated using the pseudopotential [7] to $\sim 1 \times 10^4$ atoms.

3.2. LASER IONIZATION STUDIES

First tests for the set-up and the suitability of the Ti:Sa laser system were carried out at the RISIKO isotope separator without the repelling electrode or trap. Stable isotopes of eight different elements (Al, Ca, Ni, Ga, Ge, Sn, Gd and Yb) were investigated so far. At laser ion beam intensities of up to 700 nA, e.g., in Sn, the non-resonant background accounted to about 10 pA mainly caused by electronic noise in the Faraday cup amplifier. Samples were inserted in metallic form into a tantalum surface ion source and evaporated at moderate temperatures of about 900°C. All step resonant excitation and ionization schemes for up to 10 elements, including Tc and Pu, have now been developed for Ti:Sa laser use and a wide range of spectroscopic studies on high lying Rydberg and autoionizing states have been carried out. These data are presently analyzed.

In addition to the spectroscopic studies, the general applicability of the Ti:Sa laser system for RILIS operation has been demonstrated during measuring runs at various off-line isotope separators in different facilities. In a demonstration at the ISAC facility at TRIUMF, Canada, selectivity increase by surface ion suppression with a repelling potential could be shown for the first time [11]. At the ISOLDE off-line isotope separator, a comparison of emittances for laser ion beams and surface ion beams was carried out, confirming the estimated numbers discussed above [12]. A final operational test in respect to reliability of the Ti:Sa laser system and high efficiency in the laser ionization has been carried out recently at the Oak Ridge National Laboratory as a step towards preparation for the future RIA on-line facility [13].

3.3. FIRST LIST TESTS

First operational tests of the laser ion source trap LIST have been carried out using the trap as an ion guide. Ca surface ions were transmitted through the RFQ in this mode without significant losses. Laser ion production with the LIST was tested with Ga, where efficient excitation and ionization via Rydberg states with only two Ti:Sa lasers is possible. In transmission mode without the repeller or bunching the ionization rate was unexpectedly suppressed by about a factor of $\sim$10. Laser ions were produced by addressing a Rydberg state and by direct non-resonant ionization into the continuum. This observed loss of efficiency is not understood up to now, as no difference in behaviour between surface and laser ions is expected in the case of laser ionization taking place inside the atomizer. Using the repeller electrode and limiting the laser ionization volume to the trap a still more significant loss in ion signal intensity of a factor of 100 compared to laser ionization without the trap was observed under similar conditions. This also

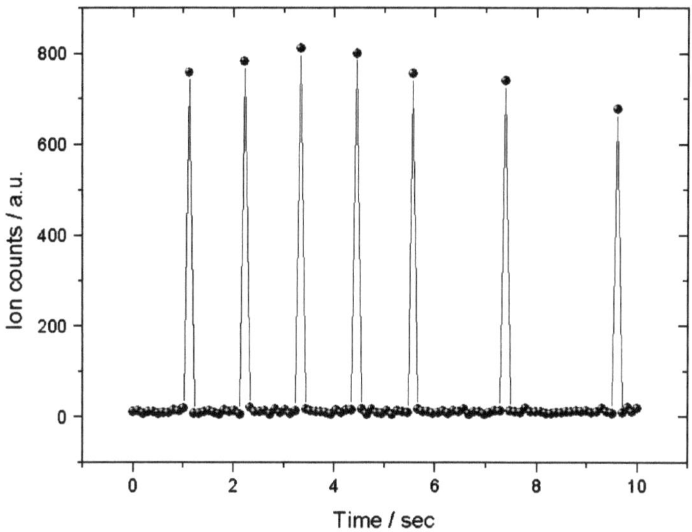

Figure 7. LIST ion bunches: ion current versus time for slow pulsing of the trap at ~1 Hz.

is about a factor of at least 10 worse than predicted from the simulation studies. Both results lead to the conclusion, that potentials in the trap do not yet match the optimum operational mode. Cleaning, realignment, further optimization of working conditions and conclusive measurements are in progress. To study the temporal behaviour the release of ion bunches was investigated using a channeltron single ion detector. Low repetition rates in the 1 Hz range were chosen together with low helium (He) pressures of 1×10^{-4} mbar. Bunched ion beams are shown in Figure 7, due to still missing synchronization to the laser pulse the appearance is not yet regular. Temporal structure of individual ion bunches and studies at realistic repetition rates and He pressures will be carried out after installation of a fast detector and data acquisition system which is presently in progress.

4. Conclusion

The novel laser ion source trap LIST, as first suggested in [2], is under development at Mainz University, presently undergoing first tests and characterization measurements. As a first step, a dedicated Ti:Sa laser system has been set up for these activities, exhibiting very promising specifications for RILIS use. During measurements at various off-line isotope separators, this solid state laser system has proven its applicability and reliability. Off-line ionization efficiencies comparable to the ones obtained in on-line experiments with well-established dye laser systems like at the ISOLDE RILIS at CERN, have been demonstrated, e.g., in Sn [12, 13]. Identical copies of the laser system were meanwhile set up for on-line application at JYFL, Finland as well at TRIUMF, Canada. Ti:Sa resonance ionization has been successfully demonstrated for up to 10 elements so far.

The nine-fold segmented RFQ trap of the LIST with two repelling electrodes has been constructed based on extensive computer calculations. After finishing the construction of the required electronic components, the trap presently undergoes performance tests in combination with a graphite atomizer. Operation in rf-only transmission mode with and without gas supply has been demonstrated and first ion bunches at low repetition rates have been generated. Optimizations of experimental conditions are still required. We expect a successful demonstration of the full cooling and trapping capabilities within the next weeks to month, afterwards installation at on-line facilities will be considered.

Acknowledgement

We gratefully appreciate all scientific and organizational contributions from our collaborators at TRIUMF/Canada, CERN/Switzerland and Oak Ridge National Laboratory/USA. The LIST project is funded by the German BMBF under grant Mz 179.

References

1. Köster U., Fedoseyev V. N. and Mishin V. I., *Spectrochim. Acta* **58B** (2003), 1047.
2. Blaum K., Geppert C., Kluge H.-J., Mukherjee M., Schwarz S. and Wendt K., *Nucl. Instrum. Methods Phys. Res. B* **204** (2003), 331.
3. Yi J., Geppert C., Horn R. and Wendt K., *Jpn. J. Appl. Phys.* **42** (2003), 5066.
4. Horn R., Kratz J.-V., Rauth C. and Wendt K., Opt. Comm., to be published.
5. Ravn H. L. and Allardyce B. W., On-line mass separators. *Treatise on heavy-ion science* **8** (1989), 363.
6. Wendt K., Bhowmick G. K., Bushaw B. A., Herrmann G., Kratz J.-V., Lantzsch J., Müller P., Nörtershäuser W., Otten E. W., Schwalbach R., Seibert U.-A., Trautmann N. and Waldek A., *Radiochim. Acta* **79** (1997), 183.
7. Herfurth F., Dilling J., Kellerbauer A., Bollen G., Henry S., Kluge, H.-J., Lamour E., Lunney D., Moore R. B., Scheidenberger C., Schwarz S., Sikler G. and Szerypo J., *Nucl. Instrum. Methods A* **469** (2001), 254.
8. Nieminen A., Huikari J., Jokinen A., Äystö J., Campbell P. and Cochrane E. C. A., *Nucl. Instrum. Methods A* **469** (2001), 244.
9. Dahl D. A., SIMION 3D 7.0, Ion Source Software, Idaho National Engineering Laboratory (2000).
10. Schwarz S., Bollen G., Lawton D., Neudert A., Ringle R., Schury P. and Sun T., *Nucl. Instrum. Methods B* **204** (2003), 474.
11. Rauth C., Geppert C., Horn R., Lassen J., Bricault P. and Wendt K., *Nucl. Instrum. Methods Phys. Res. B* **215** (2004), 268.
12. Brück K., Eller M., Geppert C., Lettry J., Kessler T., Wendt K., Wies K. and Wilfinger R., Appl. Phys., to be published.
13. Liu, Y., Baktasch, C., Beene, J. R., Bilheux, H. Z., Haverner, C. C., Krause, H. F., Schultz, D. R., Stracener, D. W., Vane, C. R., Wendt, K., Geppert, C., Kessler T. and Brück, K., *Nucl. Instrum. Methods B* **243** (2006), 442.

Hyperfine Interactions (2005) 162:39–43
DOI 10.1007/s10751-005-9206-0

Laser Ion Source Project at IGISOL

A. NIEMINEN[1], I. D. MOORE[2,*], J. BILLOWES[1], P. CAMPBELL[1], K. T. FLANAGAN[1], CH. GEPPERT[3], J. HUIKARI[2], A. JOKINEN[2], T. KESSLER[3], B. MARSH[1], H. PENTTILÄ[2], S. RINTA-ANTILA[2], B. TORDOFF[1], K. D. A. WENDT[3] and J. ÄYSTÖ[2]
[1]*Department of Physics and Astronomy, University of Manchester, Manchester, UK.*
[2]*Department of Physics, University of Jyväskylä, Jyväskylä, Finland;*
e-mail: iain.moore@phys.jyu.fi
[3]*Institut für Physik, University of Mainz, Mainz, Germany.*

Abstract. The application of laser ionisation is being developed for the IGISOL mass separator facility in Jyväskylä, Finland. The conceived laser ion source will have two independent pulsed laser systems based on all solid-state lasers and dye lasers for maximal coverage of ionisation schemes throughout the periodic table. A laser ion source trap, LIST, method will be pursued for optimal selectivity.

1. Introduction

The ion guide method (see review in ref. [1]) has been successfully applied for the production of exotic nuclei for over 20 years. The attractive features of the method are its speed, giving access to short-lived isotopes with half-lives below 1 ms, and its chemical non-selectivity which makes it possible to study any element in the periodic table. The range of experimental apparatus coupled to the IGISOL is diverse ranging from a collinear laser spectroscopy station to a double Penning trap [2], both applied for the study of fundamental ground state (and long-lived isomeric state) properties of exotic nuclei. In addition various decay spectroscopic tools are used to study low-lying excited states. The laser ion source project was started in order to address the issues of selectivity and efficiency that are currently lagging behind in the IGISOL performance.

The use of laser ion sources at nuclear structural facilities has been demonstrated, advanced and developed to a point where it is the favoured production mechanism at ISOLDE, CERN, and now used for over 60% of the experiments [3]. In the case of ion guide type sources, comparable to that of the new facility at the IGISOL, a careful systematic development both off-line and on-line has been achieved by the LISOL group of the University of Leuven [4]. Both groups rely on tunable dye lasers to provide the pumping and ionising laser

* Author for correspondence.

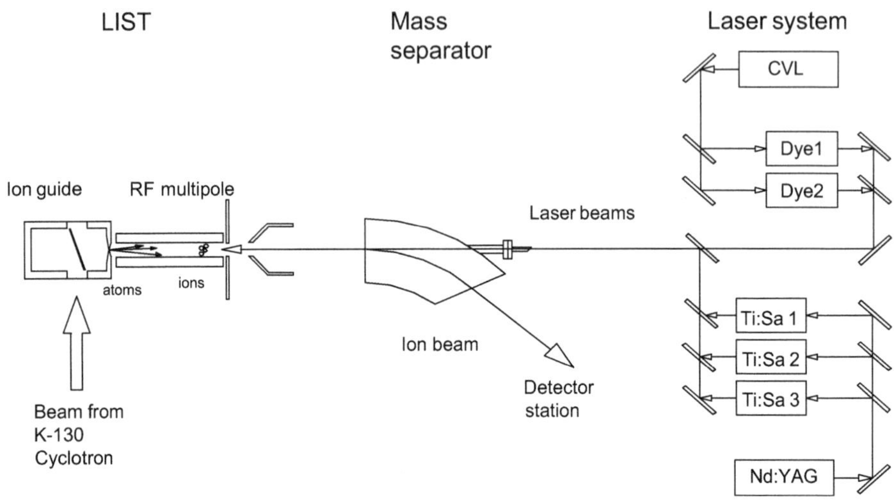

Figure 1. Laser ion source trap at IGISOL.

light. More recently, however, the TRIUMF resonant laser ion source (TRILIS) has been demonstrated using an entirely solid-state laser system, based on titanium sapphire (Ti:Sa) lasers [5] developed at the University of Mainz (Germany). Aside from the ease of using solid-state lasers, the ready production of elements with optical transition frequencies challenging for dye lasers will be achievable at the facility.

2. Experimental setup

Two very different types of ionisation geometries will be pursued at the IGISOL. Firstly ionisation in the ion guide gas volume, similar to the work done in Leuven, will be developed. The ionisation lasers can be transported to the gas cell either through the exit nozzle, through the back or through a window on the side of the ion guide. A completely different approach is the so-called LIST – Laser Ion Source Trap method [6]. Neutral atoms are allowed to exit the ion guide within the gas flow and then they are selectively re-ionised with counter-propagating laser beams. The laser ions are captured in an RF trap located immediately after the exit nozzle and guided toward the mass separator. The expected laser beam spot size is up to 6 mm, relevant in the overlap between the laser beams and the atom jet. The LIST scheme is illustrated in Figure 1.

The latter method allows a complete suppression of the primary ions by a repeller electrode placed at the exit nozzle. Critical to the efficiency of the method is the gas jet–laser beam overlap. A series of off-line tests were conducted to study the shape of the gas jet with different pumping geometries and exit nozzle shapes. Imaging of the gas jet was done by taking a photograph of the delayed light emitted from a gas discharge inside the gas cell [7]. Figure 2

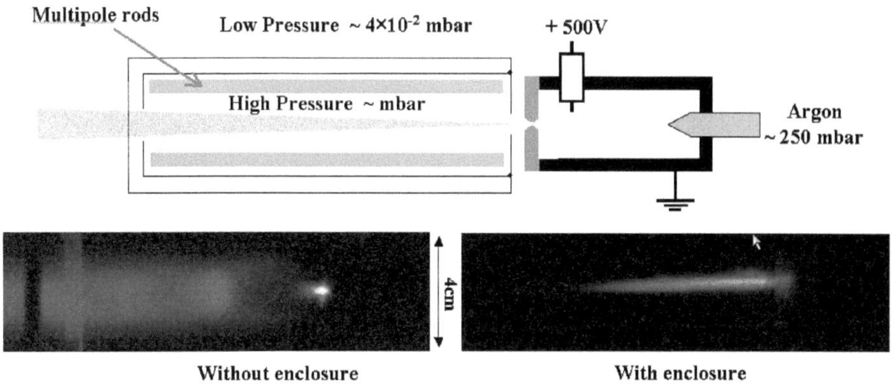

Figure 2. Illustration of the gas jet exiting the ion guide. *Top*: experimental setup. *Bottom left*: Image of the gas jet with unrestricted gas flow resulting in low background pressure. *Bottom right*: Image of the gas jet with restricted gas flow resulting in high background pressure (few mbar).

shows two cases where the pressure outside of the ion guide volume is controlled by enclosing the gas jet volume in a tube and varying the aperture at the far end of the tube. In the left picture the volume is open resembling the unperturbed gas flow. As can be seen in Figure 2 (left) the gas jet fills the 2 cm diameter tube and would result in a poor overlap between the laser beam and atom jet. In Figure 2 (right) the far end aperture is restricted to 1 cm increasing the pressure in the immediate vicinity of the ion guide exit aperture. The jet is clearly collimated to roughly 6 mm diameter over a long distance. The same effect is reproduced by changing the background pressure by restricting the pumping speed thus affecting the pressure in the whole volume outside the ion guide.

3. The lasers

Two separate high-repetition rate laser systems will provide the necessary laser frequencies to give access to ionisation schemes covering most of the periodic table. The first system is an all solid state laser system comprised of a high repetition rate Nd:YAG pump laser (Lee Laser) and three tunable titanium sapphire lasers developed and built at Mainz University. The second system is comprised of a Copper vapour laser (CVL) (Oxford lasers) pumping a pulsed dye laser and a pulsed dye amplifier seeded by a narrow band-width cw dye laser. A complete list of the lasers used in the Jyväskylä laser ion source project are shown in Tables I and II.

4. Status and outlook

As of the beginning of October 2004 the solid-state laser system has been operational at the IGISOL facility. The Nd:YAG laser is performing to the

Table I. Solid state lasers for the Jyväskylä laser ion source

Laser	Wavelength (nm)	Power (W)	Repetition rate (kHz)	Comments
Nd:YAG	532	100	1–50	Pump laser for titanium sapphires
Titanium sapphire	700–960	~2.5		Fundamental wavelength
Titanium sapphire	360–470	~0.3	As above	Frequency doubled
Titanium sapphire	240–315	~0.02		Frequency tripled
Nd:YAG	1064	5	0.02	For laser ablation

Table II. Dye lasers for the Jyväskylä laser ion source

Laser	Wavelength (nm)	Power (W)	Repetition rate (kHz)	Comments
Copper vapour	511/578	45	1–15	Pump laser for dye lasers
Pulsed dye laser	600–670		As above	Fundamental wavelength
	300–335			Frequency doubled
Pulsed dye amplifier	600–670		As above	Fundamental wavelength
	300–335			Frequency doubled
380 Ring dye laser	600–670		Continuous	For seeding the pulsed dye amplifier

factory specifications and all three Ti:Sapphire lasers are running with comparable performance to the system earlier built in Mainz [5]. Output power in the fundamental frequency is up to 2.5 W as expected with pump power of 12 W for a single Ti:Sa laser. The temporal and spatial overlap of the lasers needed prior to entering the ion guide has also been achieved. Gallium was selected as a first test case for the laser ion source. This is due to an ongoing program to measure accurate ground state branching ratios and the mass of the $N = Z$ nucleus ^{62}Ga. A known two-step laser ionisation scheme for gallium with 403 nm 1st step and <423 nm 2nd step can be achieved by frequency doubling the fundamental output of the Ti:Sa lasers. Frequency doubled light has been produced from one Ti:Sa laser using an LBO crystal. The achieved laser power of the doubled light was of the order of 100 mW. First ionisation tests were successfully done producing laser ions from natural gallium source in an atomic beam unit. In the near future the ionisation tests will be carried out in the ion guide gas cell and the construction of the LIST will be initiated. Further development will be conducted to access shorter wavelengths by frequency tripling with the Ti:Sa and dye laser system after installing the CVL pump laser before the end of 2004.

References

1. Äystö J., *Nucl. Phys.* **A693** (2001), 477.
2. Kolhinen V. S. *et al.*, *Nucl. Instr. Meth.* **B204** (2003), 502.
3. Köster U. *et al.*, *Spectrochimica Acta* **B58** (2003), 1047.
4. Kudryavstev Yu. *et al.*, *Nucl. Instr. Meth.* **B204** (2003), 336.
5. Rauth C. *et al.*, *Nucl. Instr. Meth.* **B215** (2004), 268.
6. Blaum K. *et al.*, *Nucl. Instr. Meth.* **B204** (2003), 331.
7. Rasi M. *et al.*, *Nucl. Instr. Meth.* **A378** (1996), 251.

Hyperfine Interactions (2005) 162:45–53
DOI 10.1007/s10751-005-9202-4

Laser Accelerated, High Quality Ion Beams

M. ROTH[1,*], A. BLAZEVIC[10], E. BRAMBRINK[1,2], M. GEISSEL[1,3], T. E. COWAN[4], J. FUCHS[2,4], A. KEMP[4], H. RUHL[4,5], P. AUDEBERT[2], J. COBBLE[6], J. FERNANDEZ[6], M. HEGELICH[6,7], S. LETZRING[6], K. LEDINGHAM[8], P. MCKENNA[8], R. CLARKE[9], D. NEELY[9], S. KARSCH[7,9], D. HABS[7] and J. SCHREIBER[7]

[1]*University of Technology Darmstadt, Darmstadt, Germany;*
e-mail: markus.roth@physik.tu-darmstadt.de
[2]*Laboratoire pour l'Utilisation des Lasers Intense, Palaiseau, France*
[3]*Sandia National Laboratories, Albuquerque, NM, USA*
[4]*University of Nevada, Reno, NV, USA*
[5]*Universität Bochum, Bochum, Germany*
[6]*Los Alamos National Laboratory, Los Alamos, NM, USA*
[7]*Ludwigs Maximilian Universität, München, Germany*
[8]*University of Strathclyde, Glasgow, UK*
[9]*Rutherford Appleton Laboratory, Didcot, UK*
[10]*Gesellschaft f. Schwerionenforschung, Darmstadt, Germany*

Abstract. Intense beams of protons and heavy ions have been observed in ultra-intense laser-solid interaction experiments. Thereby, a considerable fraction of the laser energy is transferred to collimated beams of energetic ions (e.g. up to 50 MeV protons; 100 MeV fluorine), which makes these beams highly interesting for various applications. Experimental results indicate very short pulse duration and an excellent beam quality, leading to beam intensities in the TW range. To characterize the beam quality and its dependence on laser parameters and target conditions we performed experiments at several high-power laser systems. We found a strong dependence on the target rear surface conditions allowing to tailor the ion beam by an appropriate target design. We also succeeded in the generation of heavy ion beams by suppressing the proton amount at the target surface.

We will present recent experimental results demonstrating a transverse beam emittance far superior to accelerator-based ion beams. Finally, we will discuss the prospect of laser-accelerated ion beams as new diagnostics in laser-solid interaction experiments. Special fields of interest are proton radiography, electric field imaging, and relativistic electron transport inside the target.

Key Words: fast ignition, high power lasers, laser generated ions.

Charged particles accelerated up to energies of several mega electron volts (MeVs) have been observed from laser plasma interaction for more than 20 years. Early experiments with long pulse (several nanoseconds) lasers incident on thin foils and wires observed protons and multiply charged ions accelerated

---

* Author for correspondence.

up to velocities of 10^9 cm/s. It was shown that these protons and carbons were present regardless of target material due to hydrocarbon contaminants inside the vacuum system [1]. The acceleration mechanism was shown to be the result of an electrostatic field due to charge separation in the underdense region of the expanding plasma. The maximum ion acceleration correlated strongly with hot electron temperature.

In contrast to the experiments using long pulse lasers, ions accelerated by ultra-intense, fs-laser systems are emitted predominantly from the rear, non-irradiated surface in a low divergent beam of excellent quality [2]. The dominant mechanism is understood as rear surface emission by the TNSA (Target Normal Sheath Acceleration) mechanism and published in detail in [3]. Briefly, relativistic electrons generated from the laser plasma interaction, having an average temperature of several MeV, envelope the target foil and form an electron plasma sheath on the rear, non-irradiated surface. The electric field in the sheath ($E_{stat} \sim kT_{hot}/e\lambda_D$, $\lambda_D = (\varepsilon_0 kT_{hot}/e^2 n_{e,hot})^{1/2}$) can reach > 10^{12} V/m. A few monolayers of atoms at the rear surface are field-ionized and accelerated normal to the surface by E_{stat} with the most energetic electrons always extending further out into the vacuum, maintaining the accelerating field as long as the electron temperature is high. Because of the accompanying electrons, the ion beam is space charge and current neutralized. So far mainly protons have been observed from the rear surface, originating from contaminating hydrocarbon and water vapour layers which coat the target. The ions form a collimated beam with an approximately exponential energy distribution with 5–6 MeV. The conversion efficiency from laser energy to ion beam energy can be quite high and efficiencies of order of 10% have already been measured [2]. Because of the dependence of the ion beam on the formation of the electron sheath, this process should also reveal information about the electron transport through the target. As will be shown below, the extreme strong, transient acceleration that takes place from a cold, initially unperturbed surface, results in the low beam emittance that may be limited only by the collisions with the comoving electrons during the acceleration. Such an acceleration process represents a new and potentially near-ideal kind of ion diode as compared either to the ion beams generated from plasma plumes, i.e., from the laser-heated turbulent plasma on the front side of the foils, or to the conventional plasma discharge ion sources used in accelerators.

1. Experiments

We have studied the influence of the target properties on laser-accelerated ion beams generated by multi-terawatt lasers. The experiments were performed using the 100 TW laser facility at Laboratoire pour l'Utilisation des Laser Intense (LULI), the 30 TW Trident laser at the Los Alamos National Laboratory, and the 1,000 TW Vulcan laser at Rutherford Appleton Laboratory. The targets were irradiated by pulses up to 5×10^{19} W/cm^2 ($\sim$ 300 fs, λ = 1.05 μm) at normal incidence. To characterize the ion beam parameters, we used absolutely calibrated ion spectrometers, Radio-chromic film (RCF), CR-39 nuclear track detectors, nuclear activation, neutron time-of-flight diagnostics, and Thomson Parabolas to detect heavy ions with respect to their charge-to-mass ratio. Stacking the RCF and CR-39 provided spatial beam profile information at different ion energies. Details of the experimental setup and the various diagnostics were published in [4].

2. Results

We performed a series of experiments to examine the properties of the laser accelerated ion beam. The targets used in those experiments have been thin metal foils. A high conductivity of the target material is important to guarantee an undisturbed transport of the electrons through the target. Details about the dependence of the ion acceleration and beam quality can be found in [4, 13]. The majority of the accelerated ions on regular targets were found to be protons, as mentioned in the introduction. Those protons, originating from surface contaminants outrun all other ion species due to their higher charge-to-mass ratio and screen the accelerating field thereby gaining most of the energy stored in the field. The energy distribution resembles an exponential spectrum up to a sharp cut-off energy, in our experiments at about 25 MeV. The angular dependence of the energy distribution was measured with two ion spectrometers, positioned at an angle of 0° and 13°, respectively. The maximum energy of about 25 MeV dropped to about 13 MeV at 13° normal to the target rear surface, consistent with the data from the RCF-stack. Details about the spectral distribution have been published in [5]. The ion beam pulse duration has been estimated to 10 ps as an upper limit due to the finite lifetime of the accelerating electrons [6], but recent measurements using accelerated heavy ions provided more accurate data, indicating individual, much shorter pulse durations for different ion species [7], well within the limits published in [6].

One of the interesting features of the laser accelerated ion beams is their directionality always normal to the rear surface. Structures at the target rear surface strongly influence the spatial distribution of the ion beam, as published in [4]. We have shown that beam tailoring by appropriate target design is possible, including defocusing and focusing of proton beams. In addition to the structure of the target, the imprint of the laser beam at the front surface can cause a deformation of the electron distribution at the rear surface, affecting the ion beam formation. In recent experiments we demonstrated that a strongly astigmatic focus can form an ion beam, where the orientation of the beam ellipse is perpendicular to the laser spot, consistent with numerical simulations based on the TNSA mechanism. An example is shown in Figure 1, where we used an astigmatic focus to generate an elliptical proton beam from a flat target.

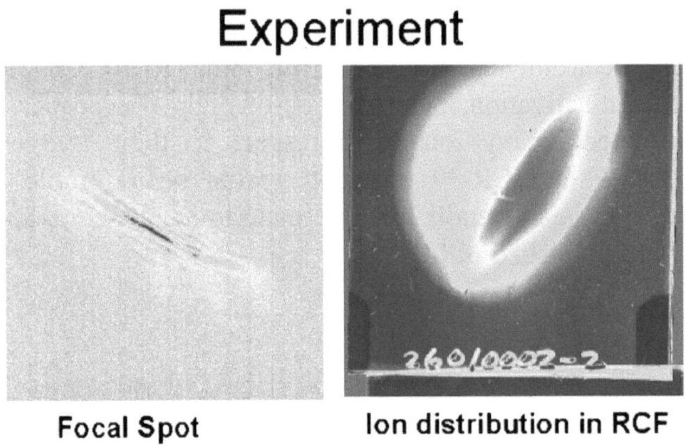

Figure 1. Effect of laser imprint on the ion beam for a strongly astigmatic laser focal spot.

For future applications of laser generated ion beams the beam quality is the most important characteristic. Especially for the use as an ion source or the application as an inertial fusion (ICF) ignitor beam [8], the ion beam emittance is crucial. The radiochromic film data suggest that the protons or other light ions accelerated by this mechanism may have a usefully small emittance in the sense of an actual ion beam.

Using a new technique that allows one to directly image the initial accelerating sheath and to fully reconstruct the transverse phase space, we have shown experimentally that, for protons of up to 10 MeV, the transverse emittance is as low as 0.004 mm mrad, i.e., 100-fold better than typical RF accelerators and at a substantially higher ion current (kA range) [9]. In addition, the removal of the comoving electrons after 1 cm of beam expansion does not increase significantly the measured proton transverse emittance. This was the first demonstration of high-current laser-produced charged-particle beams with characteristics substantially superior to conventional accelerators. Also, we determined directly for the first time the size of the ion source, a crucial parameter for developments of high-resolution charged-particle radiography or ion patterned lithography.

The technique we developed for imaging the accelerating sheath uses a target design that allows manipulating the ion beam generation during the initial, virtual cathode phase of the acceleration by generating a stream of beamlets, within the expanding proton envelope, that can be used as fiducials of the acceleration.

The laminarity of charged-particle beams is characterized by their emittance [10], which is proportional to the volume of the bounding ellipsoid of the distribution of particles in phase space. By Liouville's theorem, the phase-space volume of a particle ensemble is conserved during nondissipative acceleration and focusing. For the transverse phase-space dimensions (here $x - p_x$ for beam propagation along z), the area of the bounding phase space ellipse equals $\pi\varepsilon_N$, where the root-mean-square (rms) value of the 'normalized emittance' ε_N, at a specific beam energy (or momentum p), is expressed as $\varepsilon_N = (|p|/mc)[\langle x^2\rangle\langle x'^2\rangle -$

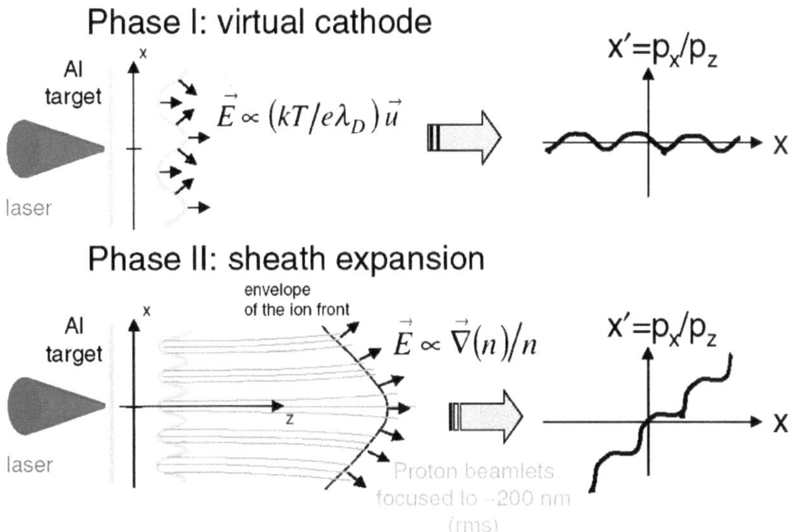

Figure 2. Schematic of experiment. The laser pulse is focused on a foil having a modulated rear surface. Protons are first accelerated normal to the surface by the virtual cathode (VC) producing a modulation of the takeoff angle. The produced beamlets then expand with the quasineutral sheath. This adds an overall near-linear divergence to the beam initial angular modulation. Projected on a film stack far away, this results in a modulation of the dose, allowing one to image the VC stage magnified.

$\langle xx'\rangle^2]^{1/2}$, where m is the ion mass, c is the velocity of light, x is the particle position within the beam envelope, and $x' = p_x / p_z$ is the particles' divergence in the x direction.

At the beam waist, we have $\varepsilon_N = \beta\gamma\sigma_x\sigma_{x'}$ where σ_x and $\sigma_{x'}$ are the rms values of the beam-width and divergence angle. For typical proton accelerators [e.g., the CERN Super Proton Synchrotron (SPS)], the emittance from the proton injector linac is ~1 mm mrad (normalized rms) and up to 3.5 mm mrad within the synchrotron, with 10^{11} protons per bunch. The longitudinal phase space ($z - p_z$) is characterized by the equivalent, energy-time product of the beam envelope and a typical value, for the CERN SPS, is ~0.5 eVs. The highest quality ion beams have the lowest values of transverse and longitudinal emittance, indicating a low effective transverse ion temperature and a high degree of angle-space and time-energy correlation.

The concept of the experiment is shown schematically in Figure 2. Laser pulses of ~20−200 J of 1 μm light (350–850 fs) were focused onto the front surface of thin foils of Au or Al (10−50 μm thick). Note that the targets can be used only once since they are destroyed during the shot.

The accelerated protons are detected in multiple layers of radiochromic film (RCF) densitometry media [11]. The spatial distribution of the protons in a given RCF layer gives the angular emission pattern at a specific interval of proton energy. By carefully preparing the rear surface of the target foil, and by shaping

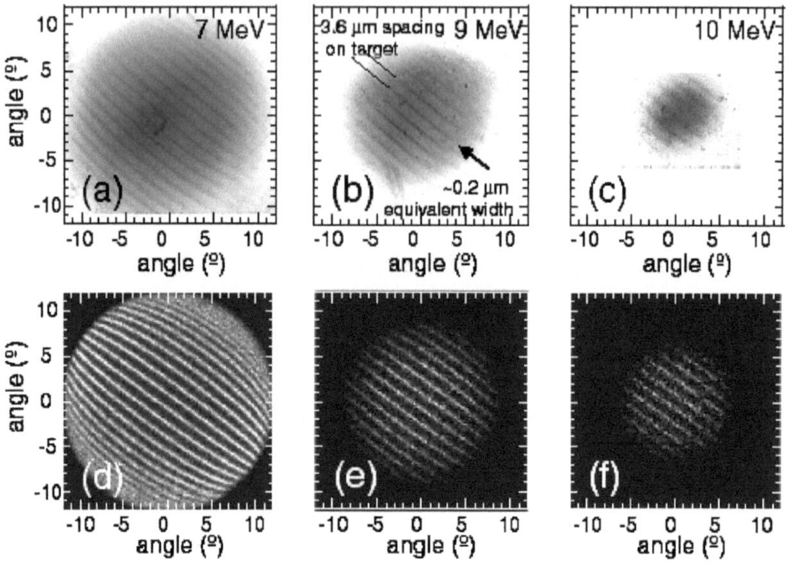

Figure 3. (a–c) Angular distributions on RCF of protons accelerated from an 18-µm thick Al flat target irradiated at 10^{19} W/cm^2. (d–f) Simulated RCF images [same parameters and proton energies as in (a–c)] using a 3D PIC effective code.

the laser focal intensity distribution, we controlled the virtual-cathode phase of the acceleration where the electric field is normal to the ion charge layer [12, 13]. For the data of Figure 3(a)–(c), we used optically flat aluminium foils on the rear surface of which we micromachined shallow grooves, 200 nm deep spaced 3.6 µm apart. From a quantitative analysis of the film optical densities, we measure that $\sim 10^{11}$ protons are produced in a single laser shot for energies above 4 MeV, which corresponds to an ion current of >1 kA at 1 mm from the target foil. We observe a decrease in the angular envelope of the protons with increasing proton energy, as has been observed previously [2, 4].

However, using the modulation of the beam intensity impressed during the initial phase, we are able to image the proton-emitting surface and thus to measure for the first time directly the source size. By accelerating protons off non-periodic surface structures we verified that there is no overlap of the beam fiducials from adjacent structures that could lead to misinterpretation of the data.

By assuming that the protons in each beamlet come from an ideal line focus, we experimentally deduce an upper limit on the transverse emittance of <0.004 mm mrad for 10 MeV protons, a factor of >100 smaller than typical proton beam sources. We attribute this to the fact that during much of the acceleration the proton space charge is neutralized by the comoving hot electrons. PIC simulations show that the image generation is more complicated.

The energy spread of the laser-accelerated proton beam is large, ranging from 0–10 MeV; however, due to the extremely short duration of the accelerating field (<10 ps), the longitudinal phase-space energy-time product must be less than

10^{-4} eVs. The 3D PIC simulations show that longitudinally the acceleration is extremely laminar, in the sense that the spread of proton energies in a given longitudinal slice is very small. We estimate from the simulation an energy-time product of $<10^{-7}$ eVs. Such extremely good longitudinal velocity 'chirp' of the beam is interesting since it could in principle allow one to monochromatize a portion of the beam by coupling it to the field gradient of a post-accelerator.

3. Heavy ion acceleration

So far mainly protons have been observed from the rear side, originating, as in the long-pulse experiments, from contaminating hydrocarbon layers which coat the targets. As soon as protons are present they outrun the heavier ions and screen the acceleration field. We present the experimental study, demonstrating that besides protons, also high-quality, high energy (~MeV/nucleon) heavy ion beams can be accelerated from the rear surface of (coated) thin foils. We find that heavy ions are effectively accelerated, provided that the hydrogenous surface contaminants are removed.

We obtained high resolution, absolutely calibrated energy spectra of different ion species, which provide additional information, not available in the proton signal, about the spatio-temporal evolution of the accelerating field and the origin of the observed ions. Details on the acceleration of heavy ions are published in [7]. To effectively remove the hydrogen contaminants we resistively heated tungsten targets up to temperatures of 1,000 K for several minutes. The ion species of interest was coated solely on the rear surface of the target, thereby unambiguously verifying the origin of the heavy ions. The proton spectrometer as well as the CR-39 did not show any protons, while strong fluorine ion tracks are observed originating from the CaF_2 layer at the target rear side. The complete removal of contaminants increased the acceleration of heavier ions considerably. Quantitative evaluation shows that F^{7+} was accelerated up to 100 MeV, i.e., more than 5 MeV/nucleon at 4% energy conversion. The RCF diagnostic confirmed this by showing a narrow spot in the first layer, which, in the absence of protons, indicates fluorine ions of energies above 4 MeV/nucleon. The evaluation of the fluorine shot shows that E-fields $E_{\mathrm{stat}} \sim 2$ TV/m on a time scale of $\tau \sim 350$ fs are necessary to accelerate F^{7+}-ions up to 100 MeV over a scale length of 1 ~10 μm. The shot presented in Figure 6 was virtually without any protons, but the modeled fields can accelerate protons up to ~25 MeV, as typical with unheated targets. The field distribution is in agreement with an extended TNSA model including dynamic fields and multiple ion species. We found that field ionization is the dominant mechanism while recombination and collisional ionization are by far less effective (see [7]). Since the targets were coated only at the back surface and we successfully removed contaminations we can rule out front side acceleration within the parameters of our experiment.

Figure 4. Radiography of a compound target. Details of the target are given in the text.

4. Applications

The excellent beam quality of the ion beam is ideally matched to the requirements for imaging techniques. One scheme of particular interest is the use of laser accelerated protons to radiograph macroscopic samples to study their properties. Due to the different interaction mechanism protons can provide complementary information to techniques like X-ray backlighting. Because of the copious amounts of protons accelerated in a very short time, laser accelerated protons provide a new diagnostic quality in research of transient phenomena.

We performed experiments to demonstrate the feasibility of these laser accelerated proton beams for radiography applications. In our experiments, we used a compound target of different materials to be imaged by the protons. It consisted of an 1 mm thick epoxy ring structure, several copper wires of 250 μm diameter, a hollow cylinder with 300 μm walls of steel, several Ti sheets of 100 μm thickness and a glass hemisphere of 900 μm diameter and 20 μm wall thickness. The protons were recorded in multiple layers of RCF to detect the image at different proton energies.

Figure 4 shows the radiography of the target for final proton energies of 7.5 MeV. The picture basically constitutes a negative image of the areal density of the target. The names of the collaborating institutes have been engraved on the epoxy ring, which results in a reduced thickness and therefore a higher energy deposition of the protons in the respective layer.

The areal density variation of the hollow cylinder, including a hole in the wall on the right hand side can be seen as well as a thin metal rod placed inside the cylinder. Close examination also shows the small glass hemisphere placed above the cylinder. The time of exposure in this experiment was estimated to be in the order of tens of picoseconds.

5. Conclusion

Laser accelerated proton and ion beams offer new prospects for a whole variety of applications. The beam quality was found to be far superior to beams accelerated by conventional accelerators with respect to the transversal beam emittance, while the longitudinal phase space was found to be comparable. The beam intensity exceeds present accelerators by orders of magnitude. The excellent beam quality implies a highly laminar acceleration and an initially extremely cold beam and the experiments proved the beam to have optical qualities. The properties are highly interesting for applications, especially as we have shown the possibility of tailoring the beam with respect to shape, ion species, efficiency, and homogeneity. A wealth of applications are currently being investigated, starting from improved diagnostic capabilities [14], to industrial and medical applications, to next generation ion sources and prospects for fast ignition in inertial confinement fusion [8, 15]. Still the underlying physics is subject to further investigation and, with respect to the extensive interest, future improvements are likely to be developed in the next years.

Acknowledgments

We acknowledge the expert support of the LULI, Trident and VULCAN laser teams. This work was supported by the EU Contract No. HPRI CT 1999-0052, and in part by grant E1127 from Region Ile-de-France, LANL Laboratory Directed Research & Development, corporate support of General Atomics and UNR Grant No. DE-FC08-01NV14050.

References

1. Gitomer S. J. *et al.*, *Phys. Fluids* **29** (1986), 2679.
2. Snavely R. *et al.*, *Phys. Rev. Lett.* **84** (2000), 640.
3. Wilks S. C. *et al.*, *Phys. Plasmas* **8** (2001), 542.
4. Roth M. *et al.*, *Phys. Rev. ST-AB* **5** (2002), 061301.
5. Allen M. *et al.*, *Phys. Plasmas* **10** (2003), 3283.
6. Hatchett S. *et al.*, *Phys. Plasmas* **7** (2000), 2076.
7. Hegelich M. *et al.*, *Phys. Rev. Lett.* **89** (2002), 85002.
8. Roth M. *et al.*, *Phys. Rev. Lett.* **86** (2001), 436.
9. Cowan T. E. *et al.*, *Phys. Rev. Lett.* **92** (2004), 204801.
10. Humphries S. (ed.), *Charged Particle Beams*, Wiley, New Jersey, 1990.
11. Klaasen N. V. *et al.*, *Med. Phys.* **24** (1997), 1924.
12. Ruhl H. *et al.*, *Phys. Plasmas* **11** (2004), L17.
13. Fuchs J. *et al.*, *Phys. Rev. Lett.* **91** (2003), 255002.
14. Borghesi M. *et al.*, *Plasma Phys. Contr. Fusion* **43** (2001), A267.
15. Temporal M. *et al.*, *Phys. Plasmas* **9** (2002), 3098.

Hyperfine Interactions (2005) 162:55–62
DOI: 10.1007/s10751-005-9203-3

PHELIX – Status and First Experiments

TH. KUEHL[1,2,*], R. BOCK[1], S. BORNEIS[1], E. BRAMBRINK[1,3], H. BRAND[1], J. CAIRD[4], E. M. CAMPBELL[5], K. CASSOU[6], E. GAUL[7], S. GOETTE[1], C. HAEFNER[1], T. HAHN[1], H. M. HEUCK[1,8], D. H. H. HOFFMANN[1,3], D. JAVARKOVA[1], A. KLISNICK[6], H.-J. KLUGE[1], S. KUNZER[1], T. MERZ[1,3], P. NEUMAYER[1,3], P. NICKLES[9], M. D. PERRY[5], D. REEMTS[1], D. ROS[6], M. ROTH[1,3], S. SAMEK[11], W. SANDNER[9], G. SCHAUMANN[1], F. SCHRADER[1], W. SEELIG[3], A. TAUSCHWITZ[1], R. THIEL[1], D. URSESCU[1,2], P. WIEWIOR[10], U. WITTROCK[8] and B. ZIELBAUER[2,9]
[1]*Gesellschaft fuer Schwerionenforschung (GSI), Darmstadt, Germany.*
[2]*Mainz University, Mainz, Germany.*
[3]*Technical University of Darmstadt, Darmstadt, Germany.*
[4]*Lawrence Livermore National Laboratory (LLNL), Livermore, USA.*
[5]*General Atomics, San Diego, USA.*
[6]*LIXAM University Paris Sud, Orsay, Cedex, France.*
[7]*University of Texas, Austin, USA.*
[8]*FH Muenster, Steinfurt, Germany.*
[9]*Max-Born-Institute, Berlin, Germany.*
[10]*FH Darmstadt, Darmstadt, Germany.*
[11]*Julius-Maximilians University of Wuerzburg, Wuerzburg, Germany.*

Abstract. The high-energy high-power laser system PHELIX (Petawatt High Energy Laser for heavy Ion eXperiments) [1] is currently under construction at the Gesellschaft fuer Schwerionenforschung mbH (GSI) Darmstadt. With PHELIX GSI will offer the unique combination of a high-current, high-energy (GeV/u) heavy-ion beam with an intense laser beam. This will open the door to a variety of fundamental science issues in the field of atomic physics, plasma physics and nuclear physics. The project will gain further interest in the near future by the dramatic increase of the accelerator performance with the starting FAIR project at GSI [2]. This paper reports the current status of the project as well as the laser architecture. The proposed physics program and a first experiment carried out with PHELIX, the realization of a transient collisionally excited x-ray laser [3], will also be reviewed briefly.

1. Introduction

An innovative research potential is opened for several relevant physics areas by the synergy through the combination of laser and accelerator beams. As a special feature of the proposed combination, the role of plasma formation, plasma

* Author for correspondence.

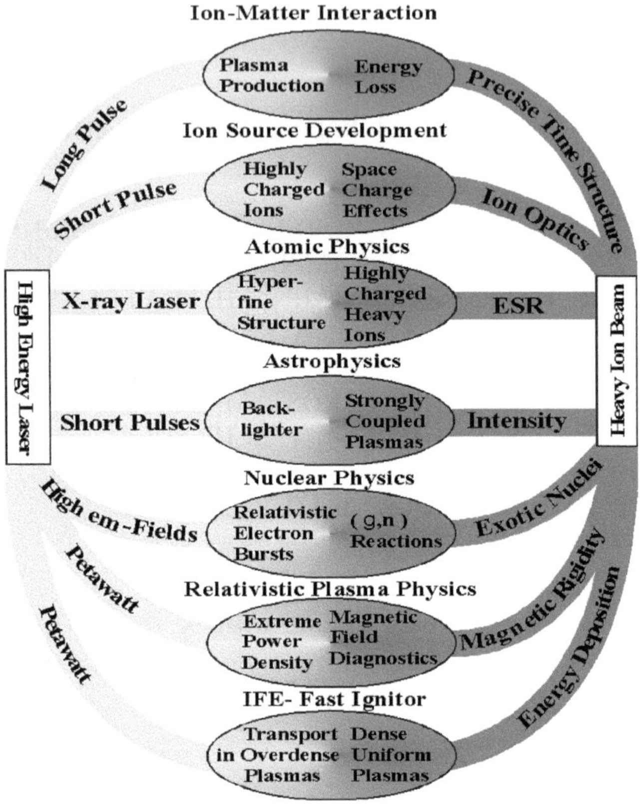

Figure 1. Physics with heavy-ion and laser beams.

interaction and diagnostics may be taken by either the ion or the laser pulses. The main task of the ion beam is to generate uniform plasma volumes as a basis for precision hydrodynamics and quantitative interaction experiments. The emphasis is on well defined geometry, small gradients, uniform ionization and high density (close to solid state density) of these plasma volumes. The laser, apart from its role as a diagnostic tool, provides ultra-short pulses for shock wave generation and has the advantage of producing much higher temperatures in surfaces or hohlraums than the ion beams at the intensities presently available. Obviously, the experimental techniques for investigations in areas discussed below will be greatly extended by such a combined laser/ion beam facility, not to mention the great discovery potential of the unique combination of a state-of-the-art high-power laser with the highest-intensity heavy-ion beams up to uranium.

A topic that touches on the plasma physics tradition already developed at GSI is the interaction of ion beams with dense, laser-generated plasmas. Besides fundamental interest in basic research these investigations of the ion-matter interaction will be aimed at evaluating the possibilities of heavy-ion beams as a driver for inertial fusion energy.

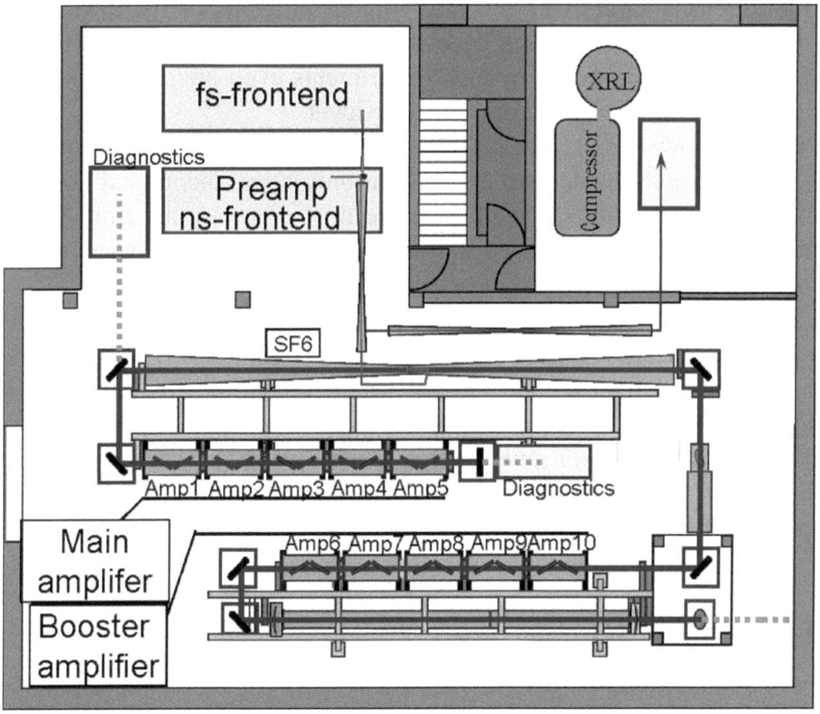

Figure 2. Layout of PHELIX.

Of interest for the further development of ion accelerators, the high-power laser can provide large numbers of highly charged heavy ions of a specific charge state at relatively low temperature. This can be utilized in novel ion sources needed in the context of possible high-intensity applications for secondary-particle production, material processing, or a fusion driver.

Atomic physics with highly charged ions, the investigation of ionization and recombination phenomena with intense laser beams is not only new physics but will also contribute significantly to the further development of x-ray lasers. Such lasers enable precision spectroscopy of few-electron ions stored in the experimental storage ring ESR. These capabilities will greatly improve current experimental possibilities of the GSI Atomic Physics program. Fundamental astrophysical questions can be accessed by producing uniform samples of dense, strongly coupled plasmas with the heavy-ion beam, using the laser as a diagnostics tool. Such plasmas are of particular interest because this largely unknown state of matter is probably found in the interior of large planets or brown dwarfs. Opacity measurements in dense plasmas will provide a data base for radiation transport in dense stellar matter and inertial fusion targets. Precision compression experiments with lasers and ion beams are aimed at determining the relevant equation of state and the first demonstration of phase transitions to new quantum states, such as metallic hydrogen (Figure 1).

High-intensity lasers can generate copious amounts of multi-MeV electrons and photons which can be used to drive nuclear reactions. The high intensities of nuclear radiation accessible in such experiments even nourish the hope to open the road to non-linear optics on a nuclear level.

Petawatt power will give access to the still mostly unexploited field of relativistic plasma physics, an area where only some theoretical concepts exist and almost no experiments were possible up to now. Apart from fundamental scientific interest, it is expected that new concepts for laser-based electron and ion acceleration will evolve.

Last but not least the combination, rather than competition, of the two most developed prospective driver media – laser beams and heavy-ion beams – might open new perspectives to the inertial fusion scenarios. An intriguing concept could be the use of a heavy-ion beam to provide improved directivity to the propagation of the laser pulse. This might prove to be inevitable for the application of ultra-short fast-ignitor pulses through the surrounding debris into the pre-compressed fusion material.

2. Laser architecture

The schematics of the PHELIX laser architecture [4] is shown in Figure 2. The desired parameters of the laser system are laser pulses of 5 kJ in long (nanosecond) pulses or alternatively pulses of ultra-high peak power (>1 PW) for focused intensities in excess of 10^{21} W/cm^2. Therefore the system uses independent front-end lasers, a so-called nanosecond front-end for pulses of 1–20 ns duration and a femtosecond front-end to provide temporally stretched pulses for chirped pulse amplification (CPA). The wavelength of these lasers is 1054 nm, as required by the pre- and main-amplifier sections, which use Nd:glass as the amplifying medium. At a common energy level of 50 mJ the pulses of either front-end are amplified in a pre-amplifier section to ~10 J and thereafter in the main-amplifier to about 1 kJ (nanosecond pulses) or 650 J (CPA pulses), respectively. Nanosecond pulses can further be amplified in a booster amplifier to the maximum energy of 5 kJ, CPA pulses will be re-compressed to 0.5 ps thus reaching peak powers on the order of 1 PW.

In the short-pulse front-end low-energy pulses of 150 fs duration are stretched in an all-reflective pulse stretcher to a duration of >2 ns and amplified by two Ti:Sapphire regenerative amplifiers at a repetition rate of 10 Hz to the energy level of 50 mJ.

A large part of the nanosecond front-end is entirely fiber based which makes it a very compact system fitting in two 19″-chassis without further need for alignment. A very attractive feature is the possibility to generate arbitrary temporal shapes with rise times of less than 100 ps by means of a fiber based amplitude modulator section in combination with an arbitrary waveform generator. Finally, a Nd:glass ring regenerative amplifier amplifies the pulses

Figure 3. Double-pass section of the main amplifier (*left*) and capacitor bank (*right*).

to the energy level of 50 mJ required for further amplification by the pre-amplifier module.

In the pre-amplifier section, single pulses from the front-end lasers are amplified in a single passage through two 19 mm and one 45 mm Nd:glass rod, pumped by flash lamps. Vacuum telescopes are used for magnification of the beam, relay imaging and spatial filtering. Large aperture Pockels cells and Faraday isolators are used for pulse cleaning and isolation against back reflections. At the exit of the pre-amplifier the pulses reach energy of up to 10 J with an approximate super-Gaussian beam profile and a 70 mm diameter. Adaptive optics is applied to maintain near diffraction-limit beam quality despite the thermal distortions in the pre-amplifiers, and also to pre-correct the anticipated effects in the main-amplifier chain. For this purpose, a closed loop active mirror system developed at the Laboratory of Adaptive Optics of the Russian Academy of Sciences is integrated at the end of the pre-amplifier.

The main-amplifier section is based on 31.5 cm aperture flash lamp pumped Nd:glass disc amplifiers, which are set up to form two groups of each five amplifiers. The first group of amplifiers is used in a double-pass (Figure 3). The pulse from the pre-amplifier is injected through a pinhole in the centre of a 15 m long 1:1 telescope, separation of the injected and the amplified beam will be realized at the pinhole of the telescope by a slight tilt of the retro-reflecting mirror. At the output of the double pass the pulse will be amplified to energy on the order of 1 kJ. The second group of amplifiers will be used in single passage further amplifying the pulse up to 5 kJ (booster amplifier). For the realization of petawatt peak power, pulses from the short-pulse front-end will be amplified in the double pass amplifier to energy of 650 J and then re-compressed in a vacuum pulse compressor.

The entire laser system is housed inside a 20 × 20 m 2-storey building. The ground floor is a class 10,000 clean room and houses the optical setup while the second floor comprises high voltage switching and the capacitor banks; a class 100 clean room for optics cleaning and preparation and the main control room.

For the combined laser/ion experiments the laser beam has to be transported to the experimental sites at the accelerator beam lines.

3. Status of PHELIX

The short pulse front-end is fully operational and shows an excellent pulse-to-pulse stability of 2–3% at typical output energy of 45 mJ, however, up to 60 mJ have been obtained. It is constantly being used for alignment of the laser chain, and to seed the pre-amplifier system. Pulses from the short pulse front-end have successfully been amplified in the pre-amplifier module to energies of 8 J. A low-energy vacuum pulse compressor has been built to compress pulses of up to 15 J to pulse duration of ~350 fs. It is used to verify pulse compression, for tests of the short pulse diagnostics equipment and for small-scale experiments. The system has successfully been used to perform a first experiment. For the final recompression of the high-energy pulse the main technical challenge is the damage threshold of the compression gratings. The emerging technology of multi-layer dielectric (MLD) gratings offers a significant improvement as compared to gold coated gratings. The large aperture compression gratings had been expected by the end of 2004, however delivery was delayed due to recent advancements in the technology. The delivery is expected mid of 2005.

Most of the large aperture components for the main-amplifier (31.5 cm diameter disc amplifiers, vacuum spatial filters, support structures) as well as high voltage components (capacitors, ignitrons, pulse forming networks) were obtained from the disassembled laser systems PHEBUS (CEA, France) and NOVA (LLNL, USA). After arrival of the NOVA components, careful inspection, cleaning and reassembling of the components had to be done, as well as refurbishing some of the optics. By now four amplifier heads are fully equipped. The hardware has been positioned and pre-aligned and commissioning of the double-pass amplifier will start after the delivery of the large diameter mirrors end of 2004.

A total of 340 capacitors had to be installed, about half of which are already tested and ready for firing the main-amplifier flash lamps. A LabVIEW™ based control system [4] has been developed in-house for full control of the high-voltage equipment as well as for access control and monitoring of safety and laser status and is currently being prepared to provide remote controlled and computer-aided alignment of the beam path.

The design for the first laser beam transport line to the accelerator was started in 2003. Completion of electrical installations and re-building of the ion beam

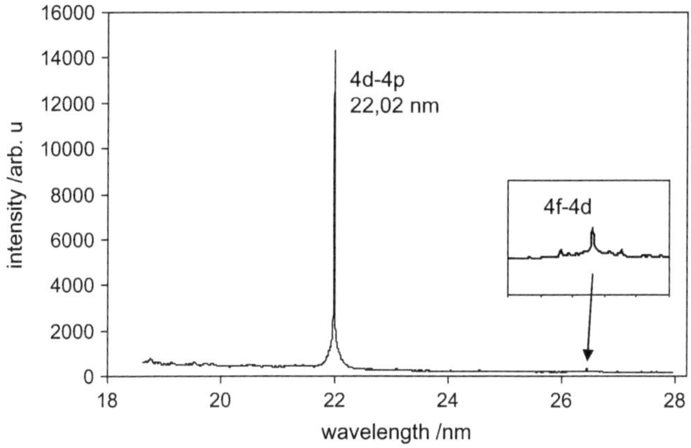

Figure 4. Spectrum obtained from the x-ray laser (Ni-like Zr) pumped with the PHELIX laser.

line are additional significant milestones on the road to the first PHELIX experiments at Z6.

4. First experiments

With the short-pulse front-end and the pre-amplifier being operational a first experiment was carried out in the test laboratory within the PHELIX building in collaboration with LIXAM, Univ. Paris Sud and MBI. In combination with the low-energy pulse compressor this sub-system was used to pump a transient collisionally excited x-ray laser. Laser pulses with an energy of 5 to 6 J generated in the pre-amplifier were used. About 20% of the pulse energy remained uncompressed and was used as a pre-pulse. This pre-pulse has to be delayed in an optical delay line to compensate for the optical pass of the main pulse in the pulse compressor. It is injected into the target chamber and focused by means of a cylindrical lens to a line focus (80 μm × 10 mm) to create the pre-plasma. The main part of the pulse was injected into the pulse compressor. After compression the pulse is transported under vacuum into the target chamber. A single on-axis parabola, under a large incidence angle of approximately 50°, was used to generate a line focus of 30 to 100 μm width over a length of 10 mm. The off-axis geometry intrinsically leads to a travelling wave excitation along the line focus with a close to the optimum speed of 1.4 times the speed of light. The delay between the pre-pulse and the compressed pulse was set to 0.7 ns. Using a zirconium target of 4.5 mm length and a total energy of 3 J on target, strong lasing was obtained from the 4d–4p transition in nickel-like zirconium at a wavelength of 22.02 nm as well as the self photo-pumped 4f–4d transition at 26.46 nm in a highly reproducible spectrum [3] shown in Figure 4. In the absence of either the pre-pulse or the short pulse, no line emission was observed. After further optimisation of the system, this x-ray laser is intended to be used to perform laser spectroscopy on highly charged heavy ions stored in GSI's Experimental Storage Ring.

5. Conclusion

As can be seen from the progress made, the PHELIX project is approaching completion. Most of the key components are tested and installed, so the system should deliver laser pulses of 700 J by the end of 2005. In addition to the x-ray laser activity a number of experiments using terawatt pulses from the pre-amplifier are proposed and will be pursued as far as allowable in this situation. Pulse compression to the petawatt level is a parallel effort to this goal. The combination of these pulses with the available heavy-ion bunches at the present GSI facility, and later at FAIR, will allow for exciting new physics.

References

1. PHELIX Project, GSI-98-10 Report, December 1998.
2. FAIR, Facility for Antiproton and Ion Research, Conceptual Design Report, GSI Report, 2003.
3. Neumayer P., Seelig W., Cassou K., Klisnick A., Ros D., Ursescu D., Kuehl Th., Borneis S., Gaul E., Geithner W., Haefner C. and Wiewior P., *Appl. Phys.* **B78** (2004), 957.
4. Bock R., Borneis S., Bruske Ch., Caird J., Habs D., Hoffmann D., Kluge H., Kühl Th., Marx D., Nickles P., Perry M., Roth M., Sandner W., Seelig W. and Tauschwitz A., In: Labaune C., Hogan W. J., Tanaka K. A. (eds.), *Inertial Fusion Sciences and Applications 99*, Elsevier, Paris, 2000, p. 703.
5. Beck D., *et al.*, *Nucl. Instr. Meth.* **A527** (2004), 567.

Hyperfine Interactions (2005) 162:63–67
DOI 10.1007/s10751-005-9205-1

Laser Spectroscopy Programme at the Jyväskylä IGISOL

J. BILLOWES
Schuster Laboratory, University of Manchester, Manchester M13 9PL, UK; e-mail: j.billowes@manchester.ac.uk

Abstract. This paper reviews the status of the laser spectroscopy programme being carried using the IGISOL mass separator in combination with an RFQ cooler-buncher. Measurements in the zirconium region are being extended to the yttrium isotopes. Two $K = 8$ isomers, in ^{176}Yb and ^{130}Ba, are found to have smaller mean square charge radii than their ground states, and the isotope shifts of stable osmium isotopes have been measured off-line by collinear laser spectroscopy.

1. Introduction

The laser spectroscopy programme at the Accelerator Laboratory, Jyväskylä (JYFL) is run by a collaboration involving the Universities of Birmingham, Jyväskylä and Manchester. The aim of the programme has been to measure nuclear moments and charge radii of radioactive nuclei. The nuclear magnetic dipole and electric quadrupole moments are obtained directly from the hyperfine structures observed on optical transitions. The difference in nuclear mean square charge radius between two isotopes is deduced from the frequency difference (isotope shift) in the optical transition. These are part per million perturbations of the atomic energy levels which are usually masked by Doppler broadening of the transition. The standard 'collinear beams' method of overlapping the laser and ion beam eliminates this problem and provides the high sensitivity required for on-line measurements of short-lived isotopes. In this way many isotopes chains have been measured at on-line isotope separators [1–3].

2. The laser/IGISOL facility

The IGISOL facility offers two significant advantages over conventional on-line mass separators: (i) the ion guide method can provide isotope beams of any element, regardless of chemistry and with fast extraction times in the region of one millisecond [4]; (ii) the RFQ cooler-buncher [5] installed in the mass-

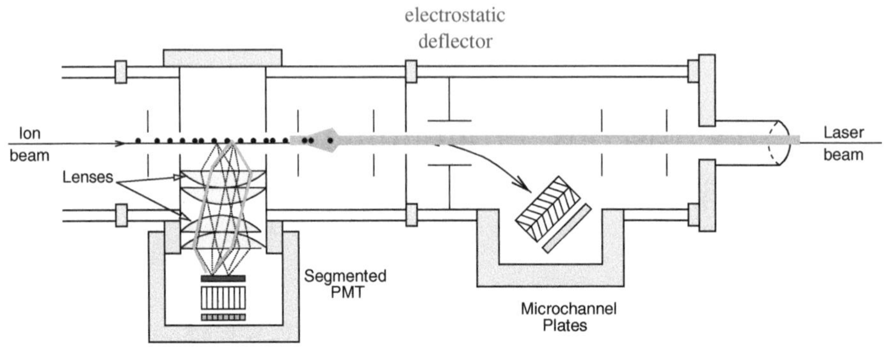

Figure 1. The *light collection region* showing an ion beam bunch passing in front of the light collection lenses.

separated line reduces the ion beam energy spread to a level where it makes a negligible contribution to the experimental resolution of the laser spectroscopy. The device can accumulate and bunch the ions before they are presented to the laser beam. A reduction in the laser-scattered light in the laser fluorescence signal of at least four orders of magnitude has been achieved [6] by accepting the fluorescence signal only while the ion bunch passes through the light collection region. The laser interaction region of the ion beam line is shown in Figure 1. Measurements have been made using this method with ion beam fluxes as low as 150 ions/s.

3. Fission fragment spectroscopy

Proton-induced fission of uranium produces a broad range of medium-mass isotopes on the neutron-rich side of stability. Using the IGISOL fission ion guide, laser measurements have been made on the zirconium isotopes out to ^{102}Zr [6]. This covers the shape transition at $N = 60$. The apparent inconsistency between the dynamic quadrupole deformations deduced from the rms charge radii and those from $B(E2; 2^+ \rightarrow 0^+)$ values for isotopes in the range 50 is still not adequately explained. A similar situation occurs in the strontium chain [7, 8]. Measurements of the quadrupole moments of ^{93,95}Zr would be useful in understanding the deformation properties here, but these isotopes are difficult to produce in a nuclear reaction. An alternative approach is to look at the odd-proton yttrium isotopes ($Z = 39$) which lie between $_{38}$Sr and $_{40}$Zr. The yttrium chain is rich in isomers and it should be possible to correlate the charge radii changes with static deformation and also see orbital-dependent effects on the core polarization.

Preparations for measuring the yttrium isotopes are well advanced. Off-line tests have been made on three ionic transitions from the $(s^2)^1S_0$ ground state. The obvious choice was the (sp) $^1P_1$ (224 nm) transition but only low laser powers are presently available and the hyperfine structure for stable ^{89}Y is unresolved. A weaker transition, (dp) $^3P_1$ (311 nm) was actually more promising because of the higher laser power available. On-line data for $^{87-89}$Y were collected showing some of the hyperfine structures, albeit on a high background. The strongest ionic transition is the (dp) $^1P_1$ (363 nm) because of (p^2) and (d^2) admixtures in the ground state and an (sp) admixture in the $^1P_1$ state. New laser optics and intra-cavity doubling crystal have been installed and the good efficiency of this transition has been experimentally confirmed. On-line running on the neutron deficient and neutron-rich isotopes will begin later in the year on this transition.

4. The K = 8 isomers in ^{130}Ba and ^{176}Yb

The celebrated ^{178m2}Hf(16$^+$) 4 quasi-particle isomer has the surprising property of a smaller rms charge radius than the ground state despite being no less deformed [9]. At JYFL it has been possible to measure two 8$^-$ isomers in ^{176}Yb and ^{130}Ba which are related to the 2-neutron and 2-proton configurations which form the 16$^+$ state. There are a number of experimental problems in these isomer shift measurements. Both $K = 8$ isomers are two-neutron configurations with small magnetic moments. Consequently, the hyperfine structure is bunched up around the much more intense nuclear ground state resonance peak. Furthermore, the quadrupole interaction can change the order of the hyperfine components and assignment of the components is difficult when some components are lost under the ground state peak. Nevertheless, an unambiguous analysis for ^{130}Ba(8$^-$) was possible [10] and, like the ^{178}Hf isomer, the state was found to have a smaller rms charge radius despite a similar deformation to the ground state.

The ^{176}Yb(8$^-$) isomer was populated in the ^{176}Yb(d,pn) reaction at 13 MeV with a deuteron beam current of 5.5 μA. The isomer component of the A=176 beam was determined by gamma-ray spectroscopy to be 200 isomers/s out of a total flux of 8,400 ions/s. The analysis has now been completed [11]. The results are compared with the measured $N = 106$ isomers in the neighbouring chains of Lu and Hf in Figure 2. For display purposes the data have been normalised to the ground states at $N = 99$ and 106 (although this is not quite consistent with the atomic factor evaluations used for the extraction of the change in mean square charge radii). It is evident from the figure that all isomers are smaller than their nuclear ground state. A comparison of the β_2 deformation parameters derived for the ground states and isomers indicate that none of the isomer shifts can be attributed to a reduction in deformation of the isomer. The reduction in rms charge radius is greatest for the 4 quasi-particle 16$^+$ state. The ^{176}Yb(8$^-$) and ^{177}Lu(23/2$^-$) states are both two-quasi-particle effects compared to their respective ground states, and are about twice the size of the normal odd-even staggering of isotope shifts which might be thought of as a one-quasi-particle effect. The most probable explanation for the isomers' smaller radii is in the

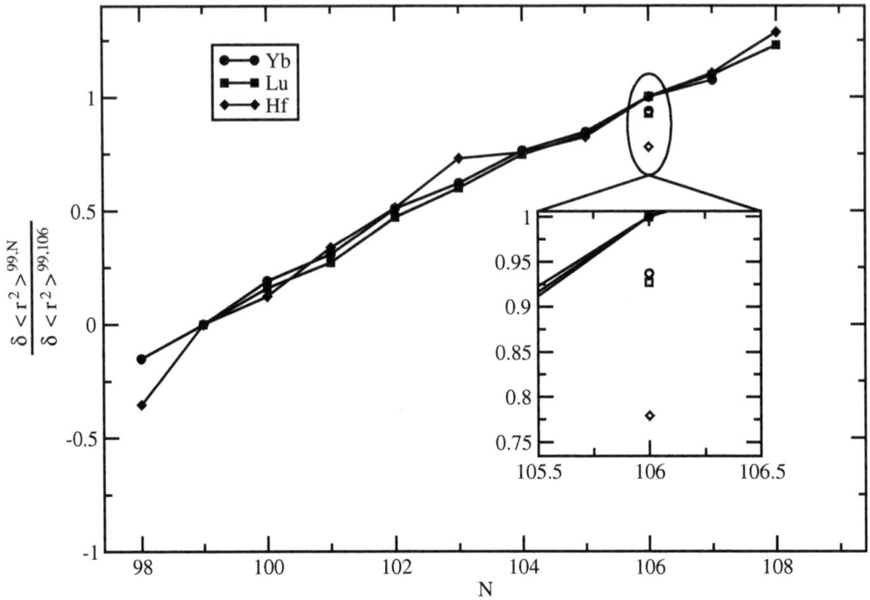

Figure 2. Relative changes in mean square charge radii for isotopes of Yb [11], Lu [12] and Hf [5, 9, 13]. The $N = 106$ isomers are shown in the panel as *open symbols.* ^{178}Hf(16^+): *diamond*; ^{177}Lu($23/2^-$): *square*; ^{176}Yb(8^-): *circle.*

pairing reduction caused by the additional quasi-particles. As yet there is no published theoretical explanation.

5. The osmium isotope chain

A programme of work has started on the osmium isotope chain with the intention of measuring the isotope shifts in the neutron-deficient region around $N = 103$. The higher-Z chains of Ir, Pt, Au and Hg all show dramatic shape changes in the vicinity of this neutron-number, and there is even a remnant of this seen for ^{175}Hf in Figure 2 above.

The ionic transition suitable for collinear spectroscopy work is the 0–43,802 cm^{-1} (9/2→7/2) transition at 228.2 nm. Off-line work has been done on the stable isotopes in order to calibrate the hyperfine factors and isotope shifts. The 228.2 nm light was generated by frequency doubling in a Coherent MBD 200 external cavity, using only 40 mW of fundamental light (456.4 nm) from a Spectra Physics 380 dye laser. The laser power in the second harmonic was calculated to be 80 μW. The fluorescence spectra for the even isotopes are shown in Figure 3. The results are consistent with atomic beam measurements on the stable isotopes (including ^{184}Os) carried out at Manchester [15] and show normal odd-even staggering with no hint of any unusual shape change down to $N = 108$. The neighbouring Ir chain has already passed through the shape

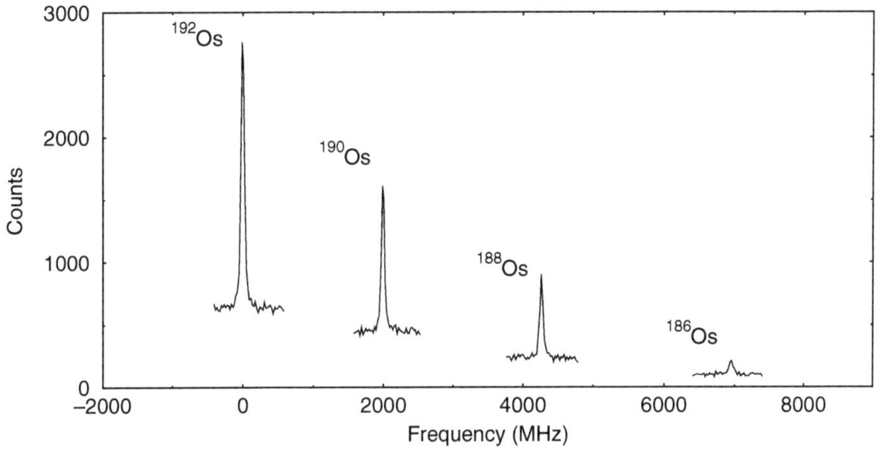

Figure 3. Laser-fluorescence spectra of the even stable Os isotopes measured by collinear spectroscopy [14].

transition by this point. Experiments are planned using the ^{185}Re(p, xn) reaction later next year.

Acknowledgements

The on-going work described in this paper is being carried out by a collaboration involving the Universities of Birmingham (G. Tungate, D.H. Forest, B. Cheal, M.D. Gardener, M. Bissel), Jyväskylä (J. Äystö, H. Penttilä, A. Jokinen, I.D. Moore, J. Huikari, S. Rinta-Antila) and Manchester (J. Billowes, P. Campbell, A. Nieminen, K. Flanagan, A. Ezwam, M. Avgoulea, B.A. Marsh and B.W. Tordoff).

References

1. Otten E. W., *Treatise on Heavy-Ion Science Vol. 8*, Plenum, New York, 1989, p. 515.
2. Billowes J. and Campbell P., *J. Phys. G* **21** (1995), 707.
3. Kluge H.-J. and Nörtershäuser W., *Spectrochim. Acta* **B58** (2003), 1031.
4. Äystö J., *Nucl. Phys. A* **693** (2001), 477.
5. Nieminen A. *et al.*, *Phys. Rev. Lett.* **88** (2002), 094801.
6. Campbell P. *et al.*, *Phys. Rev. Lett.* **89** (2002), 082501.
7. Buchinger F. *et al.*, *Phys. Rev. C* **41** (1990), 2883.
8. Lievens P. *et al.*, *Phys. Lett. B* **256** (1991), 141.
9. Boos N. *et al.*, *Phys. Rev. Lett.* **72** (1994), 2689.
10. Moore R. *et al.*, *Phys. Lett. B* **547** (2002), 200.
11. Flanagan K. T., PhD thesis, University of Manchester (2004).
12. Georg U. *et al.*, *Eur. Phys. J.* **A 3** (1998), 225.
13. Levins J. M. G. *et al.*, *Phys. Rev. Lett.* **82** (1999), 2476.
14. Avgoulea M., 1st year PhD report, University of Manchester (2003).
15. Mahgoub M. M., MSc dissertation, University of Manchester (2002).

Hyperfine Interactions (2005) 162:69–75
DOI 10.1007/s10751-005-9212-2

Resonant Ionization Laser Ion Source Project at TRIUMF

J. LASSEN[1,*], P. BRICAULT[1], M. DOMBSKY[1], J. P. LAVOIE[1,2], Ch. GEPPERT[3] and K. WENDT[3]
[1]*TRIUMF, 4004 Wesbrook Mall, Vancouver B.C., V6T 2A3, Canada;*
e-mail: lassen@triumf.ca
[2]*Laval University, Quebec City Q.C., G1K 7P4, Canada*
[3]*Johannes Gutenberg-Universität Mainz, Staudinger Weg 7, 55099 Mainz, Germany*

Abstract. Resonant laser excitation and ionisation is one of the most successful tools for the selective production of **r**adioactive **i**on **b**eams (RIB) at on-line mass separator facilities. TRIUMF plans to augment the current ion sources with a **r**esonant **i**onisation **l**aser **i**on **s**ource (RILIS), to use the high production yields from the target, as shown by the delivery of 3×10^4/s ^{11}Li ions from a standard target ion source with surface ionisation. The development and installation of TRIUMF's RILIS (TRILIS) is necessary to provide beams of short lived isotopes that conventional ion sources could not produce in sufficient intensity and purity for nuclear-, and nuclear astrophysics-experiments. A laser system consisting of three tunable **ti**tanium–**sa**pphire (TiSa) lasers with frequency doubling and tripling was employed to demonstrate first off-line resonance ionisation of Ga, and is being installed for first on-line test and a run on ^{62}Ga in December 2004.

Key Words: laser ion source, radioactive ion beams, resonance ionization, titanium–sapphire laser, TRIUMF–ISAC.

PACS: 29.25.Ni, 29.25.Rm, 32.80.Fb, 42.55.Rz, 42.60.By.

1. Introduction

TRIUMF will be able to provide up to 100 μA of 500 MeV continuous proton beam onto one of two target stations for the production of short lived isotopes from different target materials in the near future. The **i**sotope **s**eparator and **ac**celerator facility (ISAC) is built around the target-ion source to provide **r**adioactive **i**on **b**eams (RIB) for nuclear- and particle-physics experiments. Electronic documentation on the performance of different target ion source combinations is available [http://www.triumf.ca/people/marik/homepage.html] [1]. With a surface ion source high ionisation efficiencies of alkali metals, alkaline earth elements and some lanthanides are achieved, as was demonstrated

* Author for correspondence.

with the delivery of 3×10^4/s ^{11}Li ions from a standard target ion source with surface ioniser for the 'laser spectroscopic determination of the ^{11}Li charge radius.' A resonant ionisation laser ion source was identified to be the most versatile and suitable source for elements with ionisation potentials from 6 to 9 eV. The advantages are, from a practical point of view, the removal of ion source complexity from the high radiation environment of the target ion source region, and from the experimenter's view, the added element selectivity of resonant laser ionisation, which promises beams with enhanced isobar suppression.

Ion yield is always a function of all efficiencies involved from isotope production through beam delivery to the experiment. Thus for **i**sotope **s**eparator **o**n-**l**ine (ISOL) facilities the ion source always has to be viewed in context with the target chemistry, ion source and beam extraction [2]. Thus a multitude of approaches for ISOL facilities and beams exist, as is evident in the different RIB facilities currently in operation and in the planning stages around the world [3–6]. ISAC uses the continuous proton beam from the TRIUMF driver cyclotron on a thick target to produce RIB and is developing ion sources for on-line use.

The recent review of the laser ion source status quo and performance at CERN lists ionisation efficiencies and isobar suppression obtained on an element to element basis [7]. It is one of the most prominent successes of resonance ionisation spectroscopy developed in the early 1980s [8, 9]. It deserves mentioning that many of the techniques can be attributed to people involved in laser spectroscopy of short-lived isotopes [10, 11].

At TRIUMF the decision was to have an inherently reliable, low-maintenance laser system that makes use of recent improvements in solid-state laser technology/laser developments. Thus a frequency doubled, diode pumped solid-state laser was chosen as the pump source for a system of three tunable, high repetition-rate titanium–sapphire (TiSa) lasers [12] with optional frequency doubling and frequency tripling [13].

2. RILIS at TRIUMF ISAC

At TRIUMF beam development is decided on by the 'beam development committee,' which brings together the user/science group with the beam development group and laboratory management. In this way, short- and long-term planning is coordinated. The overall goal for TRILIS is to provide and develop at least one new beam each year for experiments. Current limitations are due to sharing of the off-line ion source stand and space constraints.

The TRIUMF resonant ionisation laser ion source (Figure 1) was designed to use modern, all solid-state lasers. This was done with operational considerations in mind, e.g. the maintenance requirements for the laser system under on-line operating conditions. An important reason for choosing an all solid-state laser system (over the classical Cu-vapour laser pumped dye laser system) was the

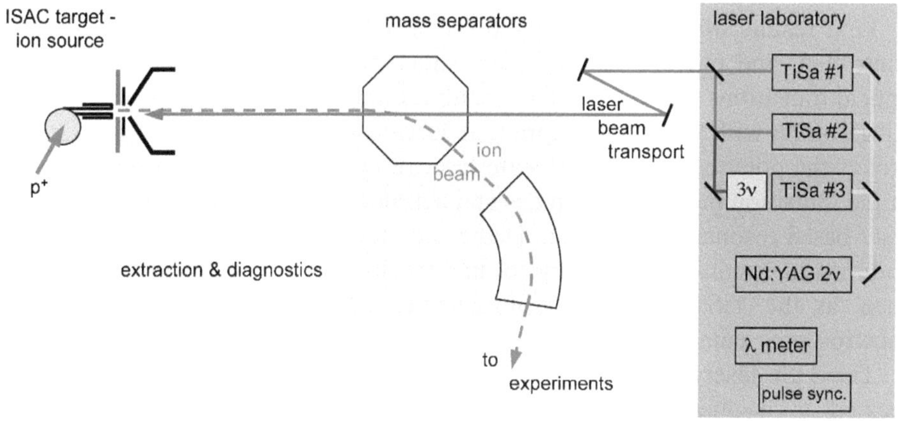

Figure 1. The TRIUMF resonant laser ion source is made up of the target ion source with a 'surface ioniser' acting as a transfer tube, the extraction ion optics, the ISAC mass separator, and the all solid-state laser system. Laser beam transport is done via dielectric mirrors.

diminishing industry support for both the Cu-vapour laser, as well as the dye-laser systems. This is especially noticeable in the availability of laser dyes.

Therefore a system of three simultaneously pumped TiSa-lasers with optional frequency doubling and tripling was constructed. The pump laser is also a high power, high repetition rate, frequency doubled YAG-laser [Coherent Corona-75, $M^2 \approx 30$, 160 ns pulse length, 5–25 kHz repetition rate, 75 W], thus adding high uptime and low maintenance requirements. Using only one, high power pump laser allows simultaneous pumping of multiple TiSa-lasers by splitting the pump laser beam. The tuneable, q-switched TiSa-lasers cover a measured wavelength range from 700–930 nm, with approximately 2 W output when pumped with 12 W at 10 kHz from the Corona pump laser. Further measurements on extending the TiSa tuning range will be carried out as necessary. The TiSa-laser pulse is about 40 ns in duration with a spectral line width of 5 GHz. This assures good spectral overlap with the atoms within the 2,000–2,200 K hot transfer tube. The TiSa laser systems require active pulse synchronisation as the pulse build up times can differ substantially. Synchronisation is done via Brewster-cut, intra-cavity Pockels-cells that are controlled from a delay gate generator. Frequency doubling efficiencies above 20% have been achieved, and overall frequency tripling efficiencies above 2% have been measured.

Unfortunately, the results from copper vapour laser pumped, dye-laser systems cannot be transferred directly, as most of these schemes rely on (i) resonant transitions that are outside the TiSa-laser tuning range, and/or rely on (ii) non-resonant, brute force photo-ionisation from the excess laser power available from the Cu-vapour pump laser (approximately 10 ns pulse length, more than 5 W at 11 kHz repetition rate at 511 and 578 nm) [7]. Pump lasers for TiSa usually are multimode with typical pulse lengths above 100 ns, so that the necessary fluences for non-resonant photo-ionisation are not available, even if the proper timing was achieved.

First results off-line [14, 15] show that TiSa laser system can be operated, controlled, and provide the necessary intensities for saturation of the selected optical transitions for resonance ionisation. Experience with ultra-trace detection RIMS on Pu with the first generation of TiSa-lasers at Mainz University [16] also proves the potential of all solid-state laser resonant ionisation. In principle most of the transition metals, lanthanides and actinide elements are accessible to TiSa-laser based resonance ionisation. Dedicated atomic spectroscopy, off-line testing and measurements are necessary for the development of each new element to be used, as the TiSa-laser excitation schemes rely on mostly unknown atomic spectroscopic data.

The TiSa-laser system with all associated diagnostics and beam transport is located in a quasi clean room environment. For successful resonant laser ionisation attention has to be paid to:

- precise wavelength control (monitoring done with 10^{-6} precision wavemeter [ATOS LM-007])
- laser beam overlap (temporal & spatial) in the ionisation region (temporal monitoring is done on fast oscilloscope, spatial monitoring done by inspection of the laser beams on a reference target. Signal optimisation is performed by optimizing the ion signal from resonance ionisation.)
- efficient laser excitation/ionisation scheme (laser excitation schemes are developed from existing spectroscopic data and tested off-line with stable isotopes. Optical spectroscopy is performed whenever spectroscopic data is insufficient.)
- saturation of optical transitions (the saturation intensity for each optical transition is determined off-line, to ensure that the transitions can be saturated under on-line conditions.)

3. Off- and on-line installations

The off-line testing with stable isotopes is done on the ISAC conditioning station (ICB), which is a copy of the on-line target ion source and exit module system. A brief description of the system can be found in the literature [14]. This duplication of the on-line system [17] ensures ready transfer of results to on-line operation.

The on-line laser laboratory has been completed and the laser beam transport to the target station is currently being installed. The principal scheme of the laser ion source system is depicted in Figure 1. The distance from the on-line lasers to the target ion source is approximately 20 m. It is planned to use the current surface ionisation source [17] directly and adding resonant laser ionisation to the surface ionisation. The surface ioniser/transfer tube connects the target to the ion extraction lens. The Ta ioniser/transfer tube is of 3 mm diameter and 30 mm length, and designed as a re-entrant heater. This transfer tube has to be kept above 2000 K in order to facilitate fast transfer and guidance of the short lived

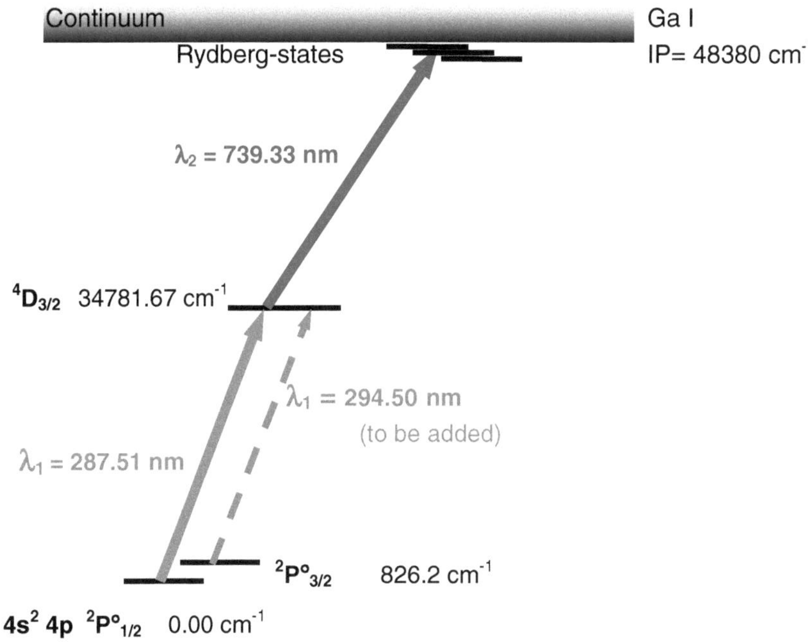

Figure 2. TiSa-laser, two-step resonant laser excitation scheme for Ga. A high lying Rydberg-state in Ga is excited. This excitation scheme relies on field-ionisation of the resonantly excited Rydberg-state. With only one frequency-tripling unit available (at present) the 287 nm first excitation will be realized.

isotopes to the extraction lens. Initial laser ion source operation will rely on the current surface ion source which will be augmented by resonant laser ionisation. It is planned to experiment with different surface materials in the transfer tube to suppress surface ionisation.

Moving the laser system from off-line to the on-line laboratory and initial beam alignment are scheduled for October 2004, with a scheduled beam time on ^{62}Ga for December 2004.

4. Test elements

The first elements to be addressed with resonant laser ionisation at TRIUMF are Ga and Al. This choice of elements is motivated by the physics program's priorities. These are ^{26}Al for nuclear astrophysics, and nuclear structure, and ^{62}Ga a super-allowed β-emitter for high precision lifetime and branching ratio measurements.

To this end a two-step resonant excitation scheme into a Rydberg-state, as shown in Figure 2, is under investigation at TRIUMF and Mainz University. A frequency tripled TiSa-laser excites and saturates the first excitation with 20 mW output power and a second TiSa-laser with 2 W output power excites the atom into a Rydberg-/autoionising-state. Pulse synchronisation is done with intracavity, Brewster-cut, Pockels-cells. Detailed investigation of Rydberg- and possibly autoionising-states will provide the most efficient second excitation step. These measurements on stable Ga are in progress at Mainz University.

5. Conclusions

The first on-line run of TRILIS on ^{62}Ga for lifetime and branching ratio measurements is scheduled for December 2004. The preparations for this first all solid-state laser based RILIS operation is ongoing and concentrate on the laser beam transport and the on-line laser lab installations at present.

Acknowledgements

The work is financed by TRIUMF which is federally funded via a contribution agreement through the National Research Council of Canada. The contributions from the lively collaboration with the research group of Dr. K.D.A. Wendt from the University of Mainz are essential to the rapid progress of TRILIS.

References

1. Dombsky M., Bricault P., Schmor P. and Lane M., *Nucl. Instrum. Methods B* **204** (2003), 191.
2. Shotter A., *Nucl. Instrum. Methods B* **204** (2003), 17.
3. Stracener D. W., *Nucl. Instrum. Methods B* **204** (2003), 42.
4. Kudryavtsev Yu. A., Facina M., Huyse M., Gentens J., Van den Bergh P. and Van Duppen P., *Nucl. Instrum. Methods B* **204** (2003), 336.
5. Köster U., *Eur. Phys. J. A* (2002), 255.
6. Van Duppen P., *Nucl. Instrum. and Methods B* **204** (2003), 9.
7. Köster U., Fedoseyev V. N. and Mishin V. I., *Spectrochim. Acta B* **58** (2003), 1047.
8. Letokhov V. S., *Nonlinear Laser Chemistry*, Springer, Berlin Heidelberg New York, 1983; and Letokhov V. S., *Laser Photoionisation Spectroscopy*, Academic Press, Orlando, 1987.
9. Hurst G. S. and Payne M. G., *Principles and Applications of Resonance Ionization Spectroscopy*, Adam Hilger, Bristol, 1989.
10. Otten E. W., In: A. Bromley (ed.), *Treatise on Heavy-Ion Physics*, Vol. 8, Plenum Press, New York, 1989.
11. Kluge H. J. and Nörtershäuser W., *Spectrochim. Acta, B* **58** (2003), 1031.
12. Horn R., Aufbau eines Systems gepulster, abstimmbarer Festkörperlaser zum Einsatz in der Resonanzionisations-Massenspektrometrie, Dissertation, Johannes Gutenberg Universität-Mainz, 2003 (in German).
13. Sirotzki S., Konzeption und Aufbau einer Frequenzverdreifachungseinheit für ein gepulstes Titan-Saphir Lasersystem, Staatsexamensarbeit, Johannes Gutenberg Universität-Mainz, 2003 (in German).
14. Rauth Ch., Geppert Ch., Horn R., Lassen J., Bricault P. and Wendt K., *Nucl. Instrum. Methods B* **215** (2004), 268.

15. Geppert Ch., Bricault P., Horn R., Lassen J., Rauth Ch. and Wendt K., *Nucl. Phys., A* **746** (2004), 631c.
16. Grüning C., Huber G., Klopp P., Kratz J. V., Kunz P., Passler G., Trautmann N., Waldeck A. and Wendt K., *Int. J. Mass Spectrom. Ion Phys.* **235** (2004), 171.
17. Dombsky M., Bishop C., Bricault P., Dale D., Hurst A., Jayamanna K., Keitel R., Olivo M., Schmor P. and Stanford G., *Rev. Sci. Instrum.* **71**(2) (2000), 978.

Hyperfine Interactions (2005) 162:77–84
DOI 10.1007/s10751-005-9210-4

Ion Mobility Measurements and Ion Chemical Reaction Studies at Heavy Elements in a Buffer Gas Cell

H. BACKE[1,*], A. DRETZKE[1], R. HORN[1], T. KOLB[1], W. LAUTH[1], R. REPNOW[3], M. SEWTZ[1] and N. TRAUTMANN[2]
[1]*Institut für Kernphysik der Johannes Gutenberg-Universität, D-55099 Mainz, Germany; e-mail: backe@kph.uni-mainz.de*
[2]*Institut für Kernchemie der Johannes Gutenberg-Universität, D-55099 Mainz, Germany*
[3]*Max-Planck-Institut für Kernphysik, D-69029 Heidelberg, Germany*

Abstract. Drift time measurements of ions in a buffer gas cell filled with argon have been performed from which changes of the ion mobility and ionic radii for various heavy elements and their compounds were determined. The ionic radius of americium shrinks by (3.1 ± 1.3)% with respect to that of plutonium, and an increase of the radius by (28 ± 2)% of plutonium oxide with respect to plutonium was found. Ion chemical reactions of erbium ions were studied online in an argon buffer gas cell to which the reaction gases oxygen (O_2) and methane (CH_4) were added. The erbium ions were implanted into the buffer gas cell with an energy of 50 MeV. The online measured reaction constant $k_{Er+O2} = (3.2 \pm 0.4) \cdot 10^{-10}$ cm^3/(molecule • s) for the reaction $Er^+ + O_2 \longrightarrow ErO^+ + O$ agrees with a reference measurement $k^{FT}_{Er+O2} = (3.6 \pm 0.4) \cdot 10^{-10}$ cm^3/(molecule • s), performed with a Fourier-Transform-Mass-Spectrometer.

Key Words: FT-ICR, heavy elements, ion chemistry, ion mobility.

1. Introduction

One of the most fascinating studies of the heaviest actinides and transactinides concerns the influence of increasing relativistic effects on the valence-electron configuration of the atoms and its consequence on chemical behavior. Relativistic effects gain particular importance for the heaviest elements in the region of $Z = 100$ so that simple extrapolation of systematics, e.g., the electron affinity, may not lead to reliable predictions any more. These relativistic effects are caused, roughly speaking, by a shrinking of the wave functions of s- and $p_{1/2}$-electrons. Inner shell electrons influence indirectly via the shielding of the nuclear potential the valence electrons as well.

---

* Author for correspondence.

It was first demonstrated in [1] that such relativistic effects for heavy elements can also be studied by a measurement of ionic radii. These can be extracted from ion mobility measurements in inert buffer gases like argon. Drift time measurements in electrical fields represent a rather simple access to the mobility of both ions and ion chemical compounds. In particular, this method is sensitive to the determination of relative changes of radii, either for simple ions as a function of the charge number Z, or for a selected element for various ion chemical compounds. From the latter, information can be extracted on the bond length, which might also be sensitive to relativistic effects.

To further investigate the reliability of this method and to get insight into the actinide shrinkage, drift time measurements of plutonium and americium have been made. These experiments and results are described in Section 3 of this contribution.

Ions of heavy elements may react in an inert buffer gas cell with small impurities as, e.g., O_2, H_2, N_2, H_2O, or CH_4. On the one hand, such reactions are unwanted in facilities like SHIPTRAP at GSI [2] where a large fractions of heavy elements should be extracted as ions and not as compounds from the buffer gas cell. On the other hand, the reactivity with the above mentioned admixtures may provide valuable information on the electronic structure of heavy elements. In preparation of such reactivity measurements we have investigated at the MP tandem accelerator facility of the MPI für Kernphysik in Heidelberg, on-line ion chemical reactions of the element erbium with admixtures of O_2 and CH_4 to the argon buffer gas. The results are compared with reference measurements performed with a Fourier-Transform-Ion-Cyclotron-Resonance (FT-ICR) spectrometer. These measurements are described in Section 4.

2. Experimental

The experimental setup is shown in Figure 1. In contrast to conventional IGISOL apparatus, where the gas jet alone transports the ions out of the cell, in this cell they drift along electric field lines created by a suitable electrode system [3]. Shortly before the ions enter the nozzle they are decoupled from the electrical field and flushed out by the gas jet. With a nozzle diameter of 1 mm and argon at a pressure of 35 mbar as a buffer gas, pumps with relatively low pumping speeds can be used (TMP I-IV). Thus, the apparatus remains rather compact. With the aid of a segmented quadrupole ion guide structure [4] and skimmers, the ions are separated from the buffer gas, mass selected in a quadrupole mass spectrometer (Balzers QMG 311) and identified by a channeltron detector.

The timing signal of the ions as obtained with a channeltron detector after resonance ionization with pulsed lasers enables a measurement of the drift time of the ions in the buffer gas cell. With the electric field known the ion mobility can be determined, which provides information on ionic radii and bond lengthes in molecules.

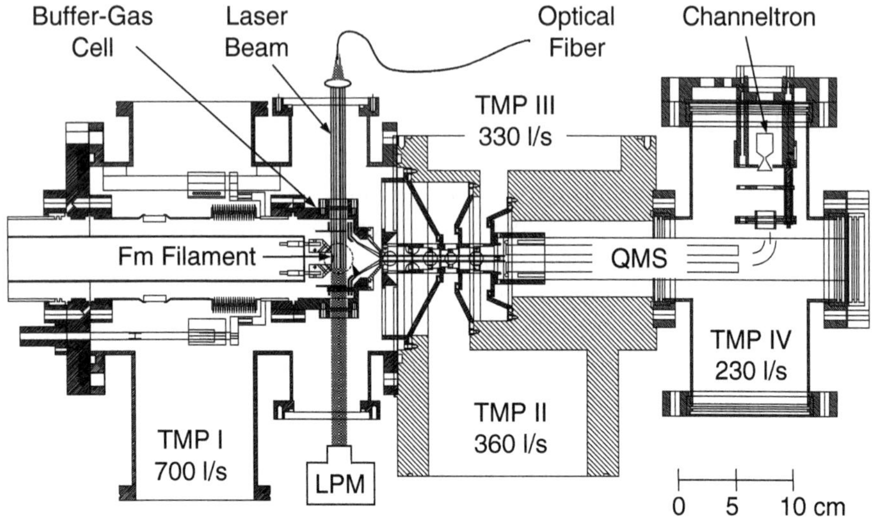

Figure 1. Experimental setup. The four sections are evacuated by Turbo Molecular Pumps (*TMP*) with relatively low pumping speeds as indicated. *QMS* is the Quadrupole Mass Spectrometer and *LPM* the Laser Power Meter.

3. Ion mobility measurements

To corroborate the effect of the actinide shrinkage as observed in our previous measurements on Cf^+ and Fm^+ [1, 5, 6], drift time measurements were performed for ^{243}Am and ^{239}Pu [7]. Am and Pu atoms were evaporated from an electrically-heated filament containing both elements, see Figure 1. Resonance ionization was accomplished with a dye laser via the 19,994 cm^{-1} and 20,426 cm^{-1} levels for americium and plutonium, respectively, and the light of the excimer laser ($\lambda = 351/353$ nm) for the second step into the continuum. The light beams were guided by two glass fibers from the lasers to the apparatus. To reduce the domain in which ions are created, the two laser beams were crossed in front of the filament under an angle of 90°. The spot where the ions are produced by resonance ionization could this way be reduced to about a diameter of 3 mm. Its distance to the filament amounted to about 2 mm. The electric field in the buffer gas cell was optimized in order to obtain long drift times of the ions at as small as possible a time spread. The drift time spectra of Am^+ and Pu^+ are depicted in Figure 2(a).

From the drift time $T = \int_S ds/v_D(s) = \frac{1}{K}\int_S ds/E(s)$ the ion mobility K can be determined. Here $v_D(s) = KE(s)$ is the ion drift velocity. If the electric field $E(s)$ along the ion trajectory S is known the ion mobility K can be determined. The latter is connected to the momentum transfer cross-section $\overline{\Omega}^{(1,1)}(T_{\mathit{eff}})$ by the expression [8]

$$K = \frac{3}{16}\frac{e}{N}\sqrt{\frac{2\pi}{\mu k T_{\mathit{eff}}}}\frac{1+\alpha}{\overline{\Omega}^{(1,1)}(T_{\mathit{eff}})} \tag{1}$$

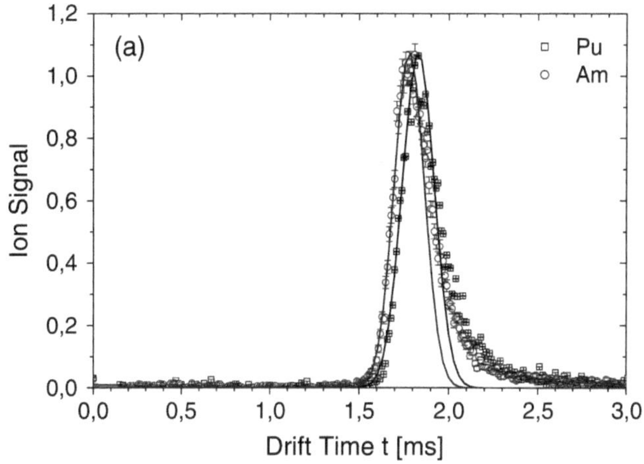

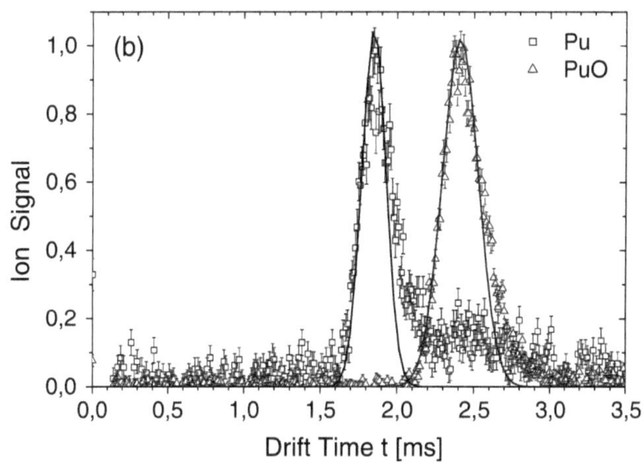

Figure 2. Drift time spectra of (a) americium (*circles*) and plutonium (*squares*) and (b) plutonium (*squares*) and plutonium oxide (*triangles*). Irradiation of laser pulses at t = 0 ms. Gaussians were fitted to the leading edges of the signals to determine the drift times.

with N = number density of buffer gas atoms, e = electric charge of the ion, μ = reduced mass, k = Boltzmann constant, T_{eff} = effective temperature, and α = higher order corrections. In the rigid sphere model, the momentum transfer cross-section can be written as

$$\overline{\Omega}^{(1,1)}(T_{eff}) = \pi \cdot d^2 = \pi(r_{Ar} + r_{ion})^2 \tag{2}$$

and, in principle, the ionic radius r_{ion} can be deduced if the radius r_{Ar} is known. Such a procedure requires the precise knowledge of a number of quantities, including the electrical field in the optical cell and the size of the argon atom.

In comparison to such an absolute measurement of ionic radii, a relative change $\Delta r^{A,B}/r^B = (r^A - r^B)/r^B$ of the ionic radii of two elements A and B can

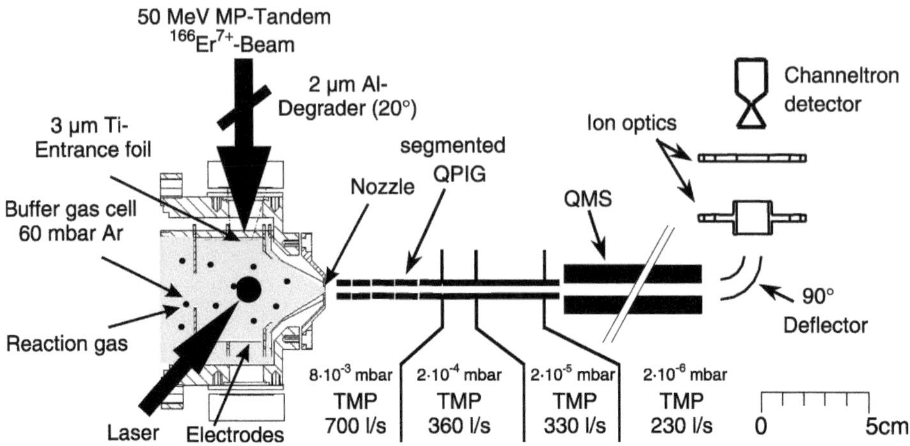

Figure 3. Experimental setup for the measurement of ion chemical reaction constants in an argon-filled buffer gas cell to which reaction gases were added.

be measured much easier and also rather precisely. From the above expressions the relative drift time difference $\Delta T^{A,B}/T^B$

$$\frac{\Delta r^{A,B}}{r^B} \approx \frac{1}{2}\frac{\Delta T^{A,B}}{T^B}\left(1+\frac{r_{Ar}}{r^B}\right) \tag{3}$$

can be derived.

A drift time difference of (0.07 ± 0.02) ms was measured between plutonium and americium, at a total drift time of (1.88 ± 0.01) ms. With these numbers equation (3) yields a relative contraction of the ionic radius of americium compared to that of plutonium of (3.1 ± 1.3)%, taking r_{Ar} = 191 pm [9] and r_{Pu} = 196 pm (calculations from Desclaux [10]). The result is in accord with the contraction of the Cf^+ relative to the Fm^+ ion, which was measured in a previous experiment [1, 5, 6]. Relativistic calculations for the differences of the *atomic* radii predict a contraction of same magnitude. Unfortunately, calculations for *ionic* radii have to the our knowledge not yet been performed.

Figure 2(b) depicts the results of drift time measurements for plutonium and plutonium oxide. From the drift times (1.85 ± 0.01) ms and (2.38 ± 0.01) ms for plutonium and plutonium oxide, respectively, the increase of the ionic radius of plutonium oxide can be determined which amounts to (28 ± 2)% [7].

Drift time measurements as those described in this section may be applied for very heavy elements which are produced with rates as low as $10^{-4}-10^{-5}$/s. Fusion reaction products come to rest in a buffer gas with very high probability as ions. In principle, a drift time measurement on a single ion is already sufficient. However, a rather sophisticated experimental setup is required for such a measurement. A feasibility study is currently underway.

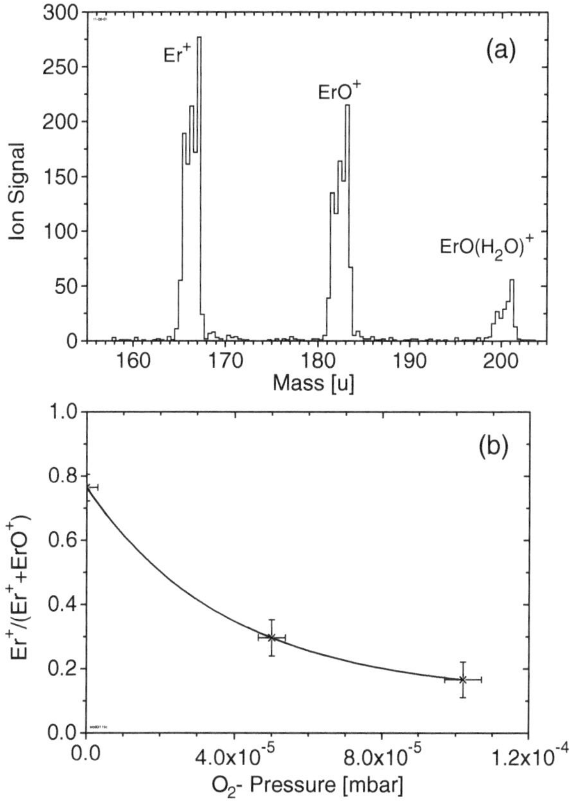

Figure 4. (a) Mass spectrum for the $Er^+ + O_2$ reaction. The $ErO(H_2O)^+$ with mass $A = 200$ is a secondary reaction product from ErO^+ with water impurities. (b) Measured ($Er^+/(Er^+ + ErO^+)$) ratio as function of the O_2 reaction gas pressure. The full line represents a best fit by an exponential function with a reaction constant $k = (3.2 \pm 0.4) \cdot 10^{-10}$ cm^3 / (molecule · s).

4. Ion chemical reaction studies on erbium

It is well known that a measurement of reaction constants for a certain chemical reaction in a group of chemical homologues provides information on the change of the electronic structure of the valence electrons [11]. Assuming that this also holds for reactions of ions in the inert gas phase with admixtures as, e.g., O_2, H_2, H_2O, N_2, or CH_4 would also provide information about the electronic structure of heavy elements as, e.g., ^{254}No or ^{256}Lr. To establish such a method, it must first be demonstrated that reaction constants can be measured in a buffer gas cell into which the ions are implanted. We have performed such measurements for the element erbium ($Z = 68$) which is the chemical homologue of fermium ($Z = 100$).

A 50 MeV Er^{7+} pulsed ion beam from the MP tandem accelerator facility of the Max-Planck Institute for Nuclear Physics (MPI-K) in Heidelberg was implanted in the buffer gas cell which was filled with 60 mbar argon, see Figure 3.

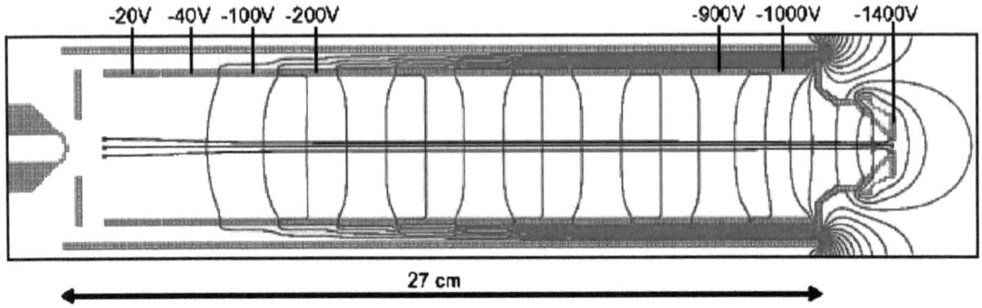

Figure 5. SIMION simulation of the electrical fields in the new drift cell. Positive ions are starting at the filament (*left side*) and drift on the shown trajectories toward the nozzle. At an argon pressure of 100 mbar and with the electrode potentials as indicated, the drift time amounts to 40 ms.

The reaction gas, oxygen (O_2) or methane (CH_4), was added to the buffer gas by glass capillaries [7]. Pressures were typically below $1 \cdot 10^{-4}$ mbar. Ions were created from the fraction of erbium which thermalized as atoms by laser resonance ionization via the level at 21,168 cm^{-1}, using a dye laser – excimer laser combination running at wavelengths of 472 nm and 351/353 nm, respectively. As in the experiments described above, the ions were guided by an electric field towards the nozzle, separated from the buffer gas jet with the aid of a segmented linear Paul trap and skimmers, mass selected in a quadrupole mass spectrometer and detected with a channeltron. A typical mass spectrum of the reaction $Er^+ + O_2 \rightarrow ErO^+ + O$ is shown in Figure 4(a).

The reaction time T is the drift time of the ions in the gas cell and amounts to $T = (2.1 \pm 0.2)$ ms. The ratio of erbium ions to the sum of ionic reaction products $Er^+/(Er^+ + ErO^+)$, as obtained from the mass spectra, is shown in Figure 4(b) as a function of the reaction gas pressure p. A reaction constant $k_{O2} = (3.2 \pm 0.4) \cdot 10^{-10}$ cm^3 / (molecule · s) was determined for the reaction $Er^+ + O_2 \rightarrow ErO^+ + O$ which is in agreement with the value $(3.6 \pm 0.3) \cdot 10^{-10}$ cm^3 / (molecule · s) as determined with the Fourier-Transform-Ion-Cyclotron-Resonance (FT-ICR) spectrometer [7].

The reaction $Er^+ + CH_4 \rightarrow ErCH_2^+ + H_2$ has not been observed. It is energetically forbidden. This result is in accord with our FT-ICR measurements [6] and also with [12]. From the mass spectra an upper limit of the reaction constant from $k_{CH4} < 2 \cdot 10^{-17}$ cm^3 / (molecule · s) was determined.

5. Conclusions and outlook

It has been demonstrated by case studies that (1) mobility measurements of ions in inert buffer gases provide information on the ionic radii, and (2) that reaction constants of ions with admixtures can be measured also in a buffer gas cell.

The ion mobility measurements have shown that the low resolution of these measurements does not allow the determination of drift time differences with an accuracy better than about 30% [7]. The main shortcomings are the initial spread of the created ions due to the finite spot size of the laser beams in combination with too short a drift region for the ions in the optical cell. To overcome this problem, a new buffer gas cell has been designed, see Figure 5. The electrode system has been optimized employing the computer code SIMION [13], and provides a rather homogeneous electrical field between filament and nozzle. With such a configuration it should be possible to determine drift time differences with an accuracy better than about 3%.

Acknowledgements

This work has been supported by Bundesministerium für Bildung und Forschung BMBF 06MZ 169I, European Union EU IONCATCHER HPRI-2001-50045, and GSI Hochschulprogramm MZ-BAC.

References

1. Backe H., Dretzke A., Eberhardt K., Fritsche S., Grüning C., Gwinner G., Haire R. G., Huber G., Kratz J. V., Kube G., Kunz P., Lassen J., Lauth W., Passler G., Repnow R., Schwalm D., Schwamb P., Sewtz M., Thörle P., Trautmann N. and Waldek A., *J. Nucl. Sci. Technol.* **Suppl. 3** (2002), 86.
2. Schönfelder J., Ackermann D., Backe H., Bollen G., Dilling J., Dretzke A., Engels O., Estermann J., Habs D., Hoffmann S., Heßberger F. P., Kluge H.-J., Lauth W., Ludolphs W., Maier M., Marx G., Moore R. B., Quint W., Rodriguez D., Sewtz M., Sikler G., Toader C. and Weber C., *Nucl. Phys.* **A 701** (2002), 579.
3. Backe H., Eberhardt K., Feldmann R., HIes M., Kunz H., Lauth W., Martin R., Schöpe H., Schwamb P., Sewtz M., Thörle P., Trautmann N. and Zauner S., *Nucl. Instrum. Methods Phys. Res.* **B 126** (1997), 406.
4. Sewtz M., Diploma Thesis, Institut für Kernphysik, Universität Mainz (1997).
5. Sewtz M., Backe H., Dretzke A., Kube G., Lauth W., Schwamb P., Eberhardt K., Grüning C., Thörle P., Trautmann N., Kunz P., Lassen J., Passler G., Dong C. Z., Fritzsche S. and Haire R. G., *Phys. Rev. Lett.* **90** (2003), 163002.
6. Backe H., Dretzke A., Fritzsche St., Haire R.G., Kolb T., Kube G., Lauth W., Ludolphs W., Morbach A., Sewtz M. and Trautmann N., *Annual Report Inst. f. Nuclear Physics 2000-2001*, 172, University of Mainz (2002).
7. Dretzke A., Doctoral Thesis, Institut für Kernphysik, Universität Mainz (2003).
8. Mason E. A. and McDaniel E. W., *Transport Properties of Ions in Gases*, Wiley, New York, 1988.
9. Laves F., *Naturwissenschaften* **25** (1937), 721.
10. Desclaux J. P., *At. Data Nucl. Data Tables* **12** (1973), 311.
11. Armentrout P. B., *Science* **251** (1991), 175.
12. Cornehl H. H., Heinemann Ch., Schröder D. and Schwarz H., *Organometallics* **14** (1995), 992.
13. Dahl A., SIMION 3D User's Manual, Idaho National Engineering and Environmental Laboratory (2000).

Hyperfine Interactions (2005) 162:85–91
DOI 10.1007/s10751-005-9208-y

Laser Spectroscopy of Transuranium Elements

YU. P. GANGRSKY*, D. V. KARAIVANOV, K. P. MARINOVA, B. N. MARKOV, YU. E. PENIONZHKEVICH and S. G. ZEMLYANOI
Joint Institute for Nuclear Research, 141980 Dubna, Russia;
e-mail: gangr@jinr.ru

Abstract. The present paper aims to discuss the prospects for nuclear structure investigation of the transuranium elements by laser spectroscopy. The authors lay stress on two peculiarities of the nuclear structure in this region: The deformed shell closure at neutron number $N = 152$ and the appearance of superdeformed isomeric states. A laser spectroscopic experimental method is proposed to study these features.

Key Words: deformed shell closure, light-induced drift, nuclear charge radii, nuclear structure, transuranium elements.

1. Nuclear structure peculiarities of the transuranium elements

The region of very high-Z nuclei ($Z > 92$) is one of the most interesting nuclear region which remains still poorly investigated. The following characteristic features of the transuranium nuclei can be pointed out:

1. Deformed shell closure with neutron number $N = 152$. The dependence of the energy of the α-decay on the neutron number with a characteristic kink at $N = 152$ [1] (see Figure 1) can be interpreted as an indication of a closed shell. This behaviour is consistent with the well-known effect at other spherical shell closures, for example $N = 126$ (see Figure 1). In addition, nuclei with $N = 152$ are the most stable isotopes toward spontaneous fission. For example, as can be seen in Figure 2, the corresponding Cm, Cf and Fm isotopes have the longest spontaneous fission half-lives.
2. Shape isomers in the $Z = 92–97$ region: U–Bk. In the nuclei of these elements isomeric states have been observed which decay predominantly by spontaneous fission (fission isomers) [3]. These states have been interpreted as lower levels in the second potential minimum of the fission barrier (Figure 3) [4]. The isomeric states have an unusually large quadrupole deformation ($\beta \approx 0.6$) which has been deduced from measurements of the rotational level lifetimes [5] and from the nuclear charge radii of the isomeric state [6].

* Author for correspondence

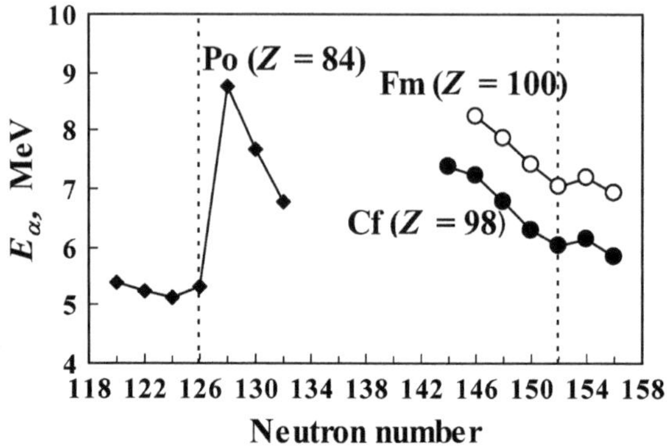

Figure 1. Dependence of the α-decay energy on the neutron number for the Po, Cf and Fm isotopes.

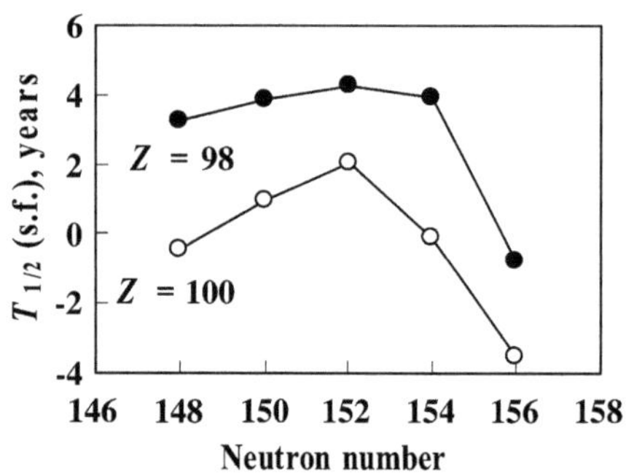

Figure 2. Spontaneous fission half-life (in logarithmic scale) *versus* neutron number for Cf and Fm isotopes around $N = 152$.

The determination of the nucleon configurations of the isomeric states and their static quadrupole moments are of particular importance. The corresponding data would be able to give more detailed insight into the structure of these states and to clarify more definitely the regions where they can occur. In a number of nuclei, e.g., in the odd U and Np isotopes (with an exception of ^{234}Np), no spontaneous fission isomers have been observed and this fact remains so far unexplained (either there are no isomers of these isotopes or the isomers exist but decay in another way).

For a rigorous and meaningful test of the nuclear structure of the transuranium elements a set of data as complete as possible should be available. This determines the steadily increased interest in studying the transuranium elements by different experimental techniques.

2. Nuclear structure investigations by laser spectroscopy

Nowadays a great deal of systematic experimental information on nuclear properties obtained by laser spectroscopy is available. The laser spectroscopic

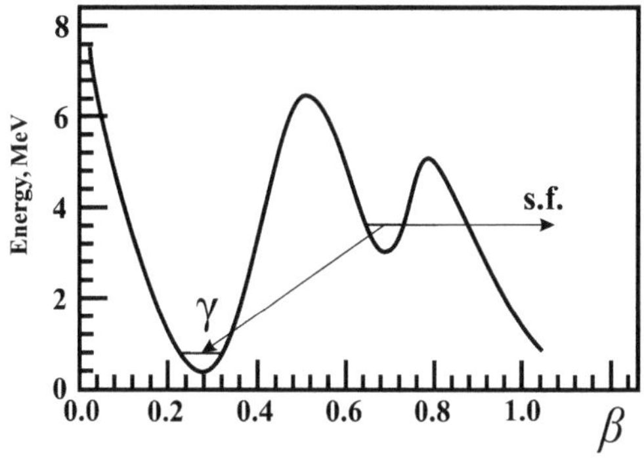

Figure 3. The nuclear potential energy in dependence on the deformation parameter.

methods are based on the electromagnetic interaction between nucleus and the electron shell. Optical isotope shifts are directly related to differences of nuclear mean square charge radii, and hyperfine structures of spectral lines contain information about nuclear spin, magnetic dipole moments and electric quadrupole moments [7]. The phenomena observed in the systematics of these quantities include collective properties (deformation, nuclear shape coexistence) and single particle properties (nucleon configurations and shell effects).

A set of such parameters have been determined for the transuranium nuclei, too. At present, data is available on the charge radii changes in isotope sequences of U [8] and Pu [9]. The charge radii development with increasing neutron number (Figure 4) is analogous to that observed in other regions of nuclei and is in agreement with the droplet model predictions. Magnetic dipole and electric quadrupole moments of several odd U, Pu and Am isotopes have also been measured [10]. As regards the fission isomers, deformation parameters, e.g. of Am, have been extracted from the isomer shifts [6]. However, this information is not sufficient for detailed explanation of the above mentioned peculiarities of the nuclear structure of the transuranium elements [2]. For example, the nuclear region around the shell closure $N = 152$ is not investigated by the laser spectroscopy methods. Thus, the necessity of further investigations is evident.

The investigation of nuclear properties of heavy elements is very difficult. A large variety of transuranium element can only be produced in heavy ion reactions with very low cross-sections, usually smaller than several millibarn. The experimental situation is very challenging because extremely high sensitivity of detection is required. For this reason new generation of high sensitive experimental methods has to be developed suitable for laser spectroscopic investigations of nuclides accessible in very low amount. For example, highly improved experimental technique is initiated already during the last years at GSI, Darmstadt. In order to search for optical transitions and to determine the nuclear ground state properties of transuranium elements, a new facility, called SHIPTRAP, is presently being build up and tested there [11].

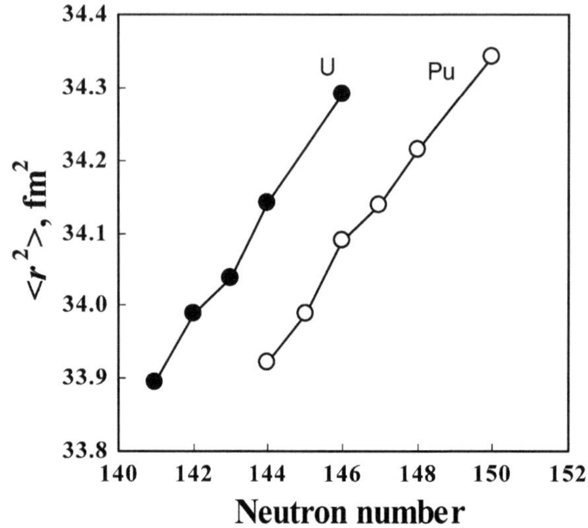

Figure 4. Dependence of the nuclear charge radii on the neutron number for the uranium and plutonium isotopes.

In the upcoming experiments, planned to be carried out by laser spectroscopy at JINR (Dubna), we intend to use the well-known method of the resonance-induced fluorescence in a collimated atomic beam as well as the method of light induced drift in a buffer gas.

3. Light induced drift of atoms in a buffer gas

An improvement of the sensitivity can be achieved if one detects radioactive decay instead of photons or ions. A way to do this may be the use of the method of light induced drift of atoms in a buffer gas [12]. The effect is based on the difference between the diffusion coefficients of excited and unexcited atoms. As a rule, excited atoms are bigger than the ground-state atoms, and for this reason their collision cross-sections with the buffer gas is larger and their diffusion coefficient is smaller. The light induced drift appears when the atoms of a selected isotope are excited resonantly, according to their velocity, by laser radiation. Due to the Doppler shift, the resonance frequencies for atoms moving in different directions will be different. The frequency shift can be expressed by the relation

$$\Delta\nu = \nu\frac{u}{c}\cos\alpha \tag{1}$$

Here u and c are the atom and light velocity, respectively; α – the angle between the directions of laser light and atom velocity. In the cases of high resolution laser spectrometers the line width is 10 to 20 times smaller than $\Delta\nu$. Thus, the

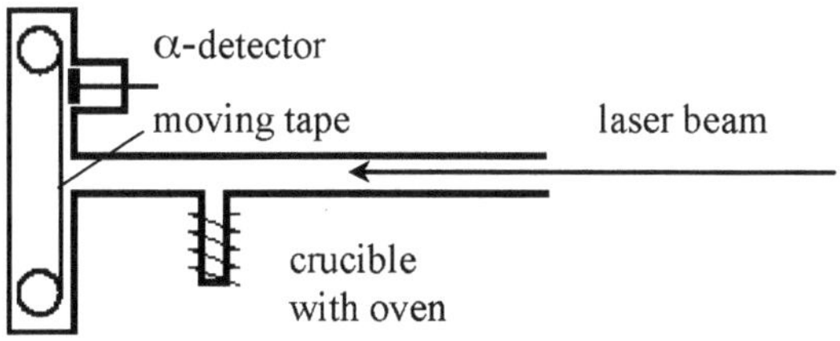

Figure 5. Principal scheme of the setup based on the laser light-induced drift.

laser light slows down the movement of the atoms in a given direction and favours their drift in an opposite direction.

The effect has already been demonstrated on the example of the radioactive isotopes ^{22}Na and ^{24}Na [13]. The isotope shift of these isotopes at the D_2 line is about 780 MHz which corresponds to Doppler shift of the resonance frequencies of atoms moving in opposite direction ($\Delta\nu = 800$ MHz at $\alpha_1 = 0°$ and $\alpha_2 = 180°$). Tuning the laser frequency in the middle of the resonance frequencies for ^{22}Na and ^{24}Na results in the drift of both isotopes in opposite directions. This has been observed experimentally in a long tube with a buffer gas (neon at a pressure of about 30 Torr) superimposed collinearly on a cw laser beam. The difference between the concentrations of ^{22}Na and ^{24}Na determined by the intensity of their γ-radiations was about two orders of magnitude.

The analogous principle can be applied for determination of the resonance frequencies of the isotopes of the transuranium elements. A light induced drift of a selected isotope can be produced taking advantage of the appropriated choice of the resonance frequency shift. Thus, the investigated isotope can be transported to the α-detector at the end of the tube and an increase of the α-counts rate will indicate the optical resonance. Of course, this implies further investigation of the optical properties of the transuranium elements to determine the changes of the atomic sizes in different excited states. As a rule, an essential increase of the atom size is distinguishing feature of the excited atomic states with a large main quantum number.

However, a realization of the light-induced drift requires multifold (more than 10^3) photon absorption by one optically active atom. This means, that a possible depopulation of the ground state via a pumping process into a metastable level has to be avoided. A usual way to do this is the inelastic scattering on the atoms of suitably chosen buffer gas (an example is the light-induced drift of the Ba atoms in nitrogen buffer gas [13]). Thus, the first experimental steps should be the choice of the buffer gas and its pressure for most effectively quenching of the intermediate metastable states.

The experimental apparatus the principle of which we propose here (Figure 5) is well suited for off line as well as for on line measurements. In the first case, a thick target could be used and the investigated element has to be chemically separated. Further the sample containing a given transuranium element is inserted in a heated crucible and the atoms evaporate into a buffer gas of the tube illuminated by laser light. In the second case, the recoils of reaction leave the target and are slowed in a buffer gas where the light induced drift can be produced. The recoils can be obtained either in reaction induced by suitable projectiles or as daughter-nuclei of alpha radioactive decay.

The method provides high detection efficiency (up to 50% for registration of α-particles) at a low background level ($<10^{-2}$ s^{-1}). Therefore, an appreciable increase of the sensitivity has to be expected. If the characteristics of the light induced drift are of the order of those obtained for sodium [14, 15] and barium [13] isotopes, experiments with transuranium elements could be performed with α-activities down to 10 s^{-1} and with sufficiently short half-lives (the drift time at distances of the order of 1 cm is $<10^{-2}$ s). It is well known, that such parameters have already been obtained in experimental setups which combine multi-step resonance ionization of atoms (or resonance neutralization of ions) and detection of radioactive decay. However, the method discussed here seems to be more easily accessible. To obtain a drift of selected isotope it is sufficiently to have only one laser frequency corresponding to a given atomic transition from the ground or low lying atomic states. This requirement is easily met. Thus, it is not necessary to use a set of tunable lasers and to know in detail the atomic level scheme up to the ionization limit. The latter is from especially importance in the case of the heavy actinide (trans-einsteinium elements) with their generally not known atomic level schemes.

4. Future experiments

At the Flerov Laboratory of Nuclear Reactions (FLNR) of JINR, Dubna an extended programme is planned for nuclear structure investigation of transuranium elements by laser spectroscopy. The programme includes study of elements heavier than plutonium. Isotopes of such elements can be produced (a) in nuclear reactors by irradiation with highly intensive neutron beams and (b) in the FLNR accelerators by heavy ion reactions. The investigated elements will be chemically separated. The laser spectroscopic studies are planned to be performed using two different methods: Laser-induced resonance fluorescence in a collimated atomic beam (in the case of long-lived isotopes) and laser light induced drift in a buffer gas (in the case of short-lived nuclei). Applying the second method one would have worse resolution, of about 0.5 GHz. Nevertheless, due to the large isotope shifts and hyperfine splitting, the nuclear parameters could be determined with a sufficient accuracy.

The first experiments will include measurements of the nuclear charge radii and nuclear moments of the Cm, Bk and Cf isotopes in the region of the neutron shell closure $N = 152$. It will be very interesting to study how the charge radii trend fits the already known behaviour around the deformed neutron shell closure. Anyway, the new results should allow one to make conclusion about the nuclear structure changes in this region.

References

1. Rytz A., *At. Data Nucl. Data Tables* **47** (1991), 205.
2. Vandenbosch R. and Huizenga J. R., *Nuclear Fission*, New York, Academic, 1973.
3. Britt H., *At. Data Nucl. Data Tables* **12** (1973), 407
4. Strutinsky V. M., *Nucl. Phys. A* **95** (1967), 420.
5. Habs D., Metag V., Specht H. and Ufert G., *Phys. Rev. Lett.* **35** (1977), 387.
6. Backe H., Hies M. and Kunzer H. *et al.*, *Phys. Rev. Lett.* **80** (1998), 920.
7. Otten E. W., *Treatise on Heavy-Ion Science* **8**, Plenum, New York, 1989, p. 515.
8. Gangrsky Yu. P., Zemlyanoi S. G., Kuldzhanov B. K. *et al.*, *Izv. AN USSR* **54** (1990), 830.
9. Aufmuth P., Heilig K. and Steudel A., *At. Data Nucl. Data Tables* **37** (1987), 445.
10. Raghavan P., *At. Data Nucl. Data Tables* **37** (1989), 189.
11. Dilling J. *et al.*, *Hyperfine Interact.* **127** (2000), 491.
12. Gelmukhanov E. Kh. and Shalagin A. M., *JETF letters* **29** (1979), 173.
13. Hradecny C., Slovak J., Tethal T. *et al.*, *Phys. Rev. A* **51** (1995), 1232.
14. Gangrsky Yu. P., Hradecny C. and Zemlyanoi S. G., *JETP* **106** (1994), 825.
15. Gangrsky Yu. P., Hradecny C., Slovak J. *et al.*, *Phys. Lett. A* **168** (1992), 230.

Hyperfine Interactions (2005) 162:93–100
DOI 10.1007/s10751-005-9211-3

First Measurement of the Nuclear Charge Radii of Short-Lived Lithium Isotopes

WILFRIED NÖRTERSHÄUSER[1,*], ANDREAS DAX[1,**], GUIDO EWALD[1], STEFAN GÖTTE[1], REINHARD KIRCHNER[1], H.-JÜRGEN KLUGE[1], THOMAS KÜHL[1], RODOLFO SANCHEZ[1], AGNIESZKA WOJTASZEK[1,†], BRUCE A. BUSHAW[2], GORDON W. F. DRAKE[3], ZONG-CHAO YAN[4] and CLAUS ZIMMERMANN[5]
[1]*Gesellschaft für Schwerionenforschung, D-64291 Darmstadt, Germany; e-mail: W.Noertershaeuser@gsi.de*
[2]*Pacific Northwest National Laboratory, P.O. Box 999, Richland, WA 99352, USA*
[3]*Department of Physics, University of Windsor, Windsor, Ontario N9B 3P4, Canada*
[4]*Department of Physics, University of New Brunswick, Fredericton, New Brunswick E3B 5A3, Canada*
[5]*Physikalisches Institut, Eberhard Karls Universität Tübingen, D-72076 Tübingen, Germany*

Abstract. A novel method for the determination of nuclear charge radii of lithium isotopes is presented. Precise laser spectroscopic measurements of the isotope shift in the lithium $2s \rightarrow 3s$ transition are combined with highly accurate atomic physics calculation of the mass dependent isotope shift to extract the charge-distribution-sensitive information. This approach has been used to determine the charge radii of ^{8,9}Li for the first time.

Key Words: halo nucleus, isotope shift, laser spectroscopy, nuclear charge radius.

1. Introduction

Nuclear properties of lithium isotopes have attracted much interest over the last decades. Particularly the discovery of the huge matter radius of ^{11}Li by Tanihata *et al.* [1] triggered many investigations on that nucleus and the picture of a two-neutron halo was established. Today, this is still a very active field, e.g., the two-neutron separation energy [2] and the spectroscopic nuclear quadrupole moment [3] have been re-measured recently with improved accuracy, and new data on the ^{11}Li decay suggests that β decay can occur inside the ^{9}Li core and one of the two halo-neutrons can survive in a halo-like configuration [4]. A very interesting topic for nuclear structure theory is the interaction between the halo-neutrons and

* Author for correspondence.
** Current address: CERN, CH-1211 Geneva 23, Switzerland.
† Current address: Institute of Physics, Swietokrzyska Academy, PL-25-406 Kielce, Poland.

the protons inside the core. It is still an open question, whether the additional neutrons influence the proton distribution inside the ^{9}Li core. It could be solved by measuring the nuclear charge radii of ^{9}Li and ^{11}Li.

Root-means square (rms) charge radii of short-lived isotopes are usually extracted from isotope shift measurements [5, 6]. To isolate the charge radius information, the mass-dependent part of the isotope shift has to be separated from the field shift contribution. For heavy nuclei the mass effect is very small and can often be neglected, while for light nuclei the situation is reversed: The mass shift is dominant and the field shift is only on the order of 10^{-4} of the total isotope shift. Moreover, the mass shift in many-electron systems cannot be calculated easily since all electron correlations must be taken into account. Until recently, such accurate calculations were only possible for helium and helium-like systems. These results allowed a precise laser-spectroscopic determination of the ^{3}He charge radius [7] and are the foundation for a determination of the charge radius of the two-neutron halo nucleus ^{6}He [8]. A few years ago, Yan and Drake [9] succeeded in extending those calculations to three-electron systems and published results for the mass-dependent isotope shift of all lithium isotopes. These are accurate enough to allow a determination of charge radii to an accuracy of a few percent, provided that the isotope shift can be measured to a relative uncertainty of better than 10^{-5}. Such an accuracy cannot be obtained using collinear fast beam spectroscopy, one of the workhorses for isotope shift measurements on unstable isotopes. Here, the precision is limited by the knowledge of the exact beam energy and typical uncertainties for isotope shifts are on the order of 1 MHz, which is as large as the expected field shift between ^{6}Li and ^{11}Li. The desired accuracy can only be reached by a Doppler-free measurement on thermal lithium atoms or by laser spectroscopy on cooled lithium atoms or ions inside an atom or ion trap. We have chosen the former and performed laser spectroscopy on the $2s \rightarrow 3s$ two-photon transition for ^{6,7,8,9}Li at the on-line mass separator at GSI Darmstadt, Germany. We report on those measurements and discuss the results, which can be used to discriminate between different nuclear models.

2. Theory

The isotope shift (IS) in an electronic transition can be divided into the mass shift (MS) and the field shift (FS)

$$\delta\nu_{\mathrm{IS}}^{AA'} = \delta\nu_{\mathrm{MS}}^{AA'} + \delta\nu_{\mathrm{FS}}^{AA'}. \tag{1}$$

The field shift in the transition is caused by the modification of the Coulomb potential that the electrons experience while they are inside the nucleus. For light

Table I. Isotope shift of the $2\,^2S_{1/2} \rightarrow 3\,^2S_{1/2}$ transition and extracted $\delta\langle r_c^2\rangle^{A,7}$ and rms charge radii r_c

	IS, MHz	MS, MHz	$\delta\langle r_c^2\rangle^{A,7}$, fm^2	r_c, fm	Reference
^{6}Li	−11,453.95(13)	−11453.01(6)	0.60(11)	2.51(4)	This work
	−11,453.734(30)		0.47(5)	2.49(4)	[18]
			0.79(25)	2.55(4)	[14]
^{7}Li		0		2.39(3)	[14]
^{8}Li	8,635.79(15)	8635.11 (4)	−0.43(11)	2.30(4)	This work
^{9}Li	15,333.14(18)	15332.02(8)	−0.72(14)	2.24(4)	This work
^{11}Li		25102.13(17)†			

† This value was calculated using a ^{11}Li mass of 11.043796(29) amu [26]. Recently, it has been remeasured with higher accuracy [2].

isotopes, the wave function of the electron can be assumed to be constant over the nuclear volume and the field shift can be expressed as

$$\delta\nu_{\mathrm{FS}}^{AA'} = -\frac{2\pi}{3} Ze^2 \Delta|\psi_e(0)|^2 \delta\langle r_c^2\rangle^{AA'} = F_i \delta\langle r_c^2\rangle^{AA'}, \tag{2}$$

where $\Delta|\psi_e(0)|^2$ is the change of the electron charge density at the nucleus between the lower and the upper state of the transition and $\delta\langle r_c^2\rangle^{AA'}$ is the change in the mean-square nuclear charge radii between the isotopes A and A'. Provided that the mass shift $\delta\nu_{\mathrm{MS}}^{AA'}$ and the electronic part F_i can be calculated to sufficient accuracy the change in the charge radius can be easily obtained according to

$$\delta\langle r_c^2\rangle^{AA'} = \frac{\delta\nu_{\mathrm{IS}}^{AA'} - \delta\nu_{\mathrm{MS}}^{AA'}}{F_i}. \tag{3}$$

The starting point for accurate mass-shift calculations is the solution of the non-relativistic Schrödinger equation including the mass polarization operator, using variational calculations in Hylleraas coordinates. The derived wave functions are then used to calculate relativistic and QED corrections. Finite nuclear mass effects are taken into account up to second order by perturbation theory. Theoretical results for the mass shift of all lithium isotopes were first published in [9] and have since been improved by including QED recoil corrections [10] and a new approach to calculating the Bethe Logarithm [11]. The most recent values are listed in Table I. The constant F_i for the $2s \rightarrow 3s$ transition was calculated to be -1.5661 MHz/fm^2. The uncertainty in the mass shift values is mainly limited by the accuracy of the relativistic recoil term.

3. Laser excitation scheme

Figure 1(A) shows the excitation and ionization scheme used. Two-photon excitation from the $2s$ ground state to the $3s$ state (2×735 nm) is followed by

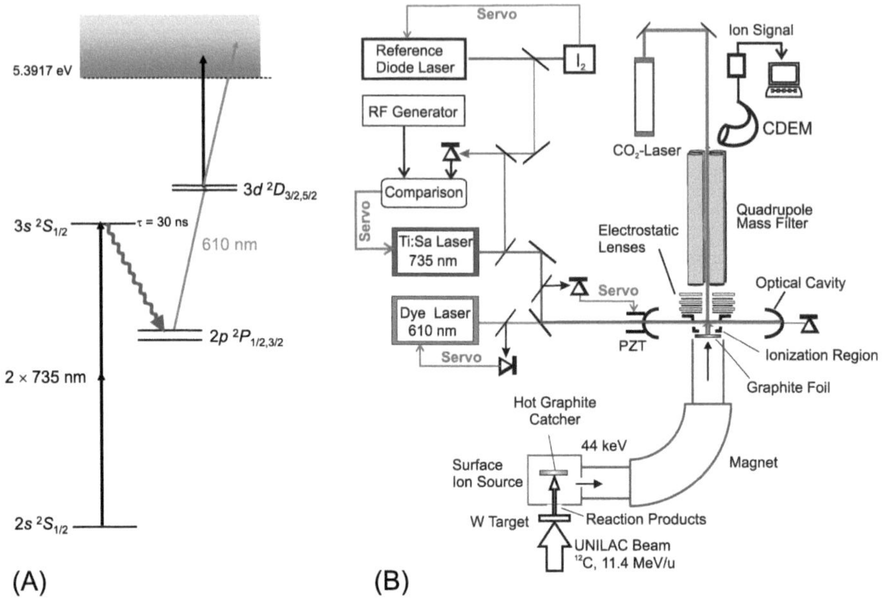

Figure 1. Excitation scheme (A) and experimental setup (B) for the resonance ionization of lithium.

resonance ionization. Ionization directly out of the 3*s* level can be attained with laser light of wavelength $\lambda < 614$ nm but is not very efficient, while resonant coupling of the 3*s* state to another, higher-lying state leads to prohibitively large AC-Stark shifts and broadening. Thus, precise spectroscopy in the $2s \rightarrow 3s$ transition must be decoupled from the efficient resonance ionization. This is achieved by including the $3s \rightarrow 2p$ spontaneous decay in the excitation scheme. Out of the 2*p* state the atoms are resonantly excited to the 3*d* state (610 nm) and then non-resonantly ionized by absorption of either a 735 nm or a 610 nm photon. Both, two-photon transition and the ionization step require high laser intensities. Therefore, both laser beams are coupled into an enhancement cavity which is built around the interaction region. This has the additional advantage that the $2p \rightarrow 3d$ transition is strongly power-broadened and a precise tuning of the 610 nm light to one of the fine structure transitions is not necessary. Rather, the 3*d* fine structure levels will overlap and provide an increased ionization efficiency. With this excitation scheme we have measured overall efficiencies exceeding 10^{-4}, which is sufficient for measurements on ^{11}Li with production yields on the order of only 10^4/s.

4. Experimental method

The experimental setup is shown schematically in Figure 1(B). The short-lived isotopes are produced by an 11.4 MeV/u ^{12}C beam impinging onto a graphite or a tungsten target for ^{8}Li and ^{9}Li, respectively. The reaction products are surface ionized, accelerated to 44 keV and mass separated in a 60° sector magnet. Typical yields of approximately 200,000 and 150,000 ions/s are obtained for ^{8}Li and ^{9}Li, respectively. The ions are then stopped inside a thin graphite foil (catcher) which is heated with a CO_2 laser to $\sim$ 2000°C. The neutralized particles diffuse out of the foil and drift into the ionization region of a quadrupole mass filter (QMS). When they cross the laser beams they are resonantly ionized, the ions are extracted with the ion optics of the QMS, mass separated and detected with a channeltron-type detector. The laser system is shown in the left part of Figure 1(B). A diode laser stabilized to the a_1 component of the I_2 $X^1\Sigma_g^+ \rightarrow BO_u^+ R(114)11-2$ hyperfine line using frequency modulation saturation spectroscopy provides a reference frequency relative to which the lithium resonance frequencies are measured. A titanium:sapphire (Ti:Sa) laser is used to produce the 735-nm light for the two-photon transition and a dye laser operated with Rhodamine 6G generates the 610-nm light for the $2p \rightarrow 3d$ transition. The Ti:Sa is stabilized relative to the reference laser using frequency-offset locking. The beat frequency between the reference laser and the Ti:Sa is counted with an RF counter and provides an accurate frequency axis for the resonance signals. The cavity is locked to the stabilized Ti:Sa laser via the Pound–Drever–Hall method while the dye laser is locked to a longitudinal mode of the cavity that is close to the resonance frequency of the $2p \rightarrow 3d$ transition. The resonators free-spectral range of $\sim$ 500 MHz and the strong saturation broadening ensures that a locking point within the flat top of the transition profile can always be found. More details of the experimental setup and the laser system can be found in [12, 13].

5. Results and discussion

To observe the resonance signals, the Ti:Sa frequency is scanned across the two-photon resonances for the different isotopes and the count rate at the detector is recorded. For isotope-shift determinations, the observed resonance profiles are fitted with a line profile consisting of a Voigt profile for the narrow two-photon component and a broad underlying Gaussian profile caused by absorption of two photons from the same direction. A spectrum of ^{9}Li is shown in Figure 2(A). Selection rules for a $^2S_{1/2} \rightarrow {}^2S_{1/2}$ two-photon transition dictate that $\Delta F = 0$. Thus, only two well-separated hyperfine components are observed. In the region between the two peaks the laser frequency has been quickly tuned without data acquisition. The symbols indicate the experimental data points and the solid line shows the fitted resonance profile. The fitted center frequencies of the hyperfine components are combined to obtain the center of gravity. This value is then corrected for residual AC Stark shifts and used to calculate the change in the mean-square charge radius $\delta\langle r_c^2\rangle^{7,A}$ relative to ^{7}Li according to Equation (3). To obtain absolute charge radii, we took the ^{7}Li rms charge radius of 2.39(3) fm

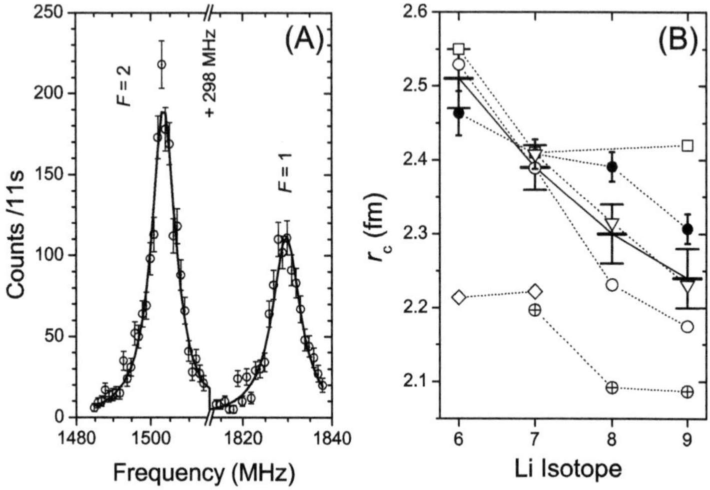

Figure 2. (A) Resonance profile of the $2s \rightarrow 3s$ two-photon transition for ^{9}Li. (B) The rms charge radii for ^{6,7,8,9}Li: (—) this measurement with $r_c(^7\text{Li})$ from electron scattering as reference; (•) obtained from interaction cross section measurements using Glauber theory [25]; (⊕) LBSM [19]; (◇) NCSM [20]; (∇) SVMC [21]; (□) DCM [22]; (∘) GFMC calculations using AV18/IL2 potentials [23, 24].

measured by electron scattering [14] as a reference. The results are listed in Table I and shown in Figure 2(B) together with predictions from different theories. The point-proton radii r_p, given in most theoretical work, have been converted to charge radii r_c by folding in the proton and neutron mean square charge radii according to

$$r_c = \sqrt{\langle r_p^2 \rangle + \langle R_p^2 \rangle + (N/Z)\langle R_n^2 \rangle}, \tag{4}$$

where $\langle R_p^2 \rangle^{1/2} = 0.895(18)$ fm [15] is the rms charge radius of the proton, $\langle R_n^2 \rangle = -0.116(5)$ fm^2 [16] is the rms charge radius of the neutron, and N and Z are the neutron and proton numbers, respectively. Note that in [13] we have used the proton radius of $\langle R_p^2 \rangle^{1/2} = 0.862$ fm [17] for the conversion. This leads to a change of about 0.01 fm for all isotopes. Five different theoretical approaches are shown in the figure: large-basis shell-model (LBSM) [19], *ab-initio* no-core shell model (NCSM) [20], stochastic variational multi-cluster (SVMC) [21], dynamic correlation model (DCM) [22] and Greens function Monte Carlo (GFMC) [23, 24] calculations. Best agreement with the experimental values is observed for the SVMC approach. GFMC calculations were carried out using a variety of effective low-energy model potentials for the two- and three-nucleon interactions. Shown are the results for the combination of the AV18 two-body with the IL2 three-body potential, because those give the best agreement with the experiment. On the other hand, the DCM, LBSM and the NSCM calculations do not agree with our results. For comparison, the figure includes model-dependent r_c values derived from experimental interaction cross-sections using Glauber-type calculations [25]. These results show a similar trend of decreasing charge radii, but to a slightly smaller extent.

The reported results demonstrate that this method is able to determine nuclear charge radii of lithium isotopes to an accuracy of a few percent, even with production yields as small as 150,000 ions/s. The goal is now the measurement of the isotope shift between ^{9}Li and ^{11}Li at the ISAC mass separator at TRIUMF (Vancouver, Canada) to determine the change in the nuclear charge radius of the ^{9}Li core when the two neutrons are added.

Acknowledgements

This work is supported from BMBF contract No. 06TU203. Support from the U.S. DOE under contract No. DE-AC06-76RLO 1830 (B.B.), NSERC and SHARCnet.(G.W.F.D. and Z.-C.Y.) is acknowledged. A.W. was supported by a Marie-Curie Fellowship of the European Community Programme IHP under contract number HPMT-CT-2000-00197.

References

1. Tanihata I., Hamagaki H., Hashimoto O., Shida Y., Yoshikawa N., Sugimoto K., Yamakawa O., Kobayashi T. and Takahashi N., *Phys. Rev. Lett.* **55** (1985), 2676–2679.
2. Bachelet C., Audi G., Gaulard C., Guénaut C., Herfurth F., Lunney D., De Saint Simon M., Thibault C. and the ISOLDE Collaboration, *Europ. Phys. J. A* **25** Supp.1 (2005), 31–32.
3. Neyens G. *et al.*, to be published.
4. Sarazin F., Al-Khalili J. S., Ball G. C., Hackman G., Walker P. M., Austin R. A. E., Eshpeter B., Finlay P., Garrett P. E., Grinyer G. F., Koopmans K. A., Kulp W. D., Leslie J. R., Melconian D., Osborne C. J., Schumaker M. A., Scraggs H. C., Schwarzenberg J., Smith M. B., Svensson C. E., Waddington J. C. and Wood J. L., *Phys. Rev. C* **70** (2004), 031302–031305.
5. Otten E. W., In: D. A. Bromley (ed.), *Treatise on Heavy-ion Science, Vol. 8*, Plenum, New York, 1989, p. 517.
6. Kluge H.-J. and Nörtershäuser W., Spectrochim, *Acta Part B* **58** (2003), 1031–1045.
7. Shiner D., Dixson R. and Vedantham V., *Phys. Rev. Lett* **74** (1995), 3553–3556.
8. Wang L.-B., Mueller P., Bailey K., Drake G. W. F., Greene, J. P., Henderson D., Holt R. J., Janssens R. V. F., Jiang C. L., Lu Z.-T., O'Connor T. P., Pardo R. C., Rehm K. E., Schiffer J. P. and Tang X. D., *Phys. Rev. Lett.* **93** (2004), 142501.
9. Yan Z.-C. and Drake G. W. F., *Phys. Rev. A* **61** (2000), 022504.
10. Yan Z.-C. and Drake G. W. F., *Phys. Rev. A* **66** (2002), 042504.
11. Yan Z.-C. and Drake G. W. F., *Phys. Rev. Lett.* **91** (2003), 113004.
12. Nörtershäuser W., Dax A., Ewald G., Katayama I., Kirchner R., Kluge H.-J., Kühl T., Sanchez R., Tanihata I., Tomaselli M., Wang H. and Zimmermann C., *Nucl. Instrum. Meth. Phys. Res. B* **204** (2003), 644–648.
13. Ewald G., Nörtershäuser W., Dax A., Götte S., Kirchner R., Kluge H.-J., Kühl T., Sanchez R., Wojtaszek A., Bushaw B. A., Drake G. W. F., Yan Z.-C. and Zimmermann C., *Phys. Rev. Lett.* **93** (2004) 113002; *Phys. Rev. Lett.* **94** (2005), 039901.
14. de Jager C. W., deVries H. and deVries C., *At. Data Nucl. Data Tables* **14** (1974), 479–508.

15. Sick I., *Phys. Lett. B* **576** (2003), 62–67.
16. Kopecky S., Harvey J. A., Hill N. W., Krenn M., Pernicka M., Riehs P. and Steiner S., *Phys. Rev. C* **56** (1997), 2229–2237.
17. Simon G. G., Schmitt C., Borkowski F. and Walther V. H., *Nucl. Phys. A* **333** (1980), 381–391.
18. Bushaw B. A., Nörtershäuser W., Ewald G., Dax A. and Drake G. W. F., *Phys. Rev. Lett.* **91** (2003), 043004.
19. Navrátil P. and Barrett B. R., *Phys. Rev. C* **57** (1998), 3119–3128.
20. Navrátil P. and Ormand W. E., *Phys. Rev. C* **68** (2003), 034305.
21. Suzuki Y., Lovas R. G. and Varga K., *Prog. Theor. Phys. Suppl* **146** (2002), 413–421.
22. Tomaselli M., Kühl T., Nörtershäuser W., Ewald G., Sanchez R., Fritzsche S. and Karshenboim S. G., *Can. J. Phys.* **80** (2002), 1347–1354.
23. Pieper S. C. and Wiringa R. B., *Annual Rev. Nucl. Part Science* **51** (2001), 53–90.
24. Pieper S. C., Varga K. and Wiringa R. B., *Phys. Rev. C* **66** (2002), 044310.
25. Tanihata I., Kobayashi T., Yamakawa O., Shimoura S., Ekuni K., Sugimoto K., Takahashi N., Shimoda T. and Sato H., *Phys. Lett. B* **206** (1988) 592–596.
26. Audi G. and Wapstra A. H., *Nucl. Phys. A* **595** (1995), 409–480.

Hyperfine Interactions (2005) 162:101–107
DOI 10.1007/s10751-005-9222-0

Charge Radii of Argon Isotopes in the $f_{7/2}$ Shell and Radii Systematics in the Ca-Region

K. BLAUM[1], W. GEITHNER[2], J. LASSEN[3], P. LIEVENS[4], K. MARINOVA[5,*], R. NEUGART[2]
[1]*Physics Department, CERN, CH 1211, Geneva 23, Switzerland*
[2]*Institut für Physik, Universität Mainz D-55099, Mainz, Germany*
[3]*TRIUMF, Vancouver BC, V6T 2A3, Canada*
[4]*Laboratorium voor Vaste-Stoffysica en Magnetisme, K.U. Leuven B-3001 Leuven, Belgium*
[5]*Laboratory of Nuclear Reaction, Joint Institute of Nuclear Research, 141980 Dubna, Russia; e-mail: marinova@nrmail.jinr.ru*

Abstract. Changes in mean square (ms) nuclear charge radii of Ar isotopes across the $1f_{7/2}$ shell are studied by fast-beam collinear laser spectroscopy using an ultra-sensitive detection method based on optical pumping and state-selective collisional ionization. The new data set on Ar, in combination with the known charge radii of K, Ca and Ti in the $\nu 1f_{7/2}$ shell, offers an opportunity to obtain a more complete overview of nuclear radii trends around the proton shell closure $Z = 20$ and between the neutron shell closures $N = 20$ and $N = 28$.

Key Words: collinear fast-beam laser spectroscopy, collisional ionization, mean square charge radii of $^{38-46}$Ar isotopes, method of direct measurement of the beam kinetic energy, nuclear radii trend in the calcium region.

1. Introduction

The unusual behaviour of the nuclear charge radii of the Ca isotopes ($Z = 20$) covering the $f_{7/2}$ neutron shell is well known [1]. It is characterized by a symmetric parabolic curve with a maximum in the middle of the shell and by the very surprising fact that the two doubly magic nuclei ^{40}Ca and ^{48}Ca have almost identical mean square charge radii [2]. However, this behaviour for the Ca isotopes is only qualitatively reproduced by the data on the neighbouring odd-Z element potassium [3]. Distinct differences include not only the mid-shell maximum compared to shell closures but even more the odd-even staggering. Moreover, earlier results for the argon isotope chain [4] covering the upper sd and partly the $f_{7/2}$ shell show a trend of generally increasing nuclear charge radii from $N = 20$ to $N = 28$. From the point of view of the gross nuclear radii systematics such large structural differences are unusual for neighbouring isotopic chains. Thus, several questions arise, for

* Author for correspondence.

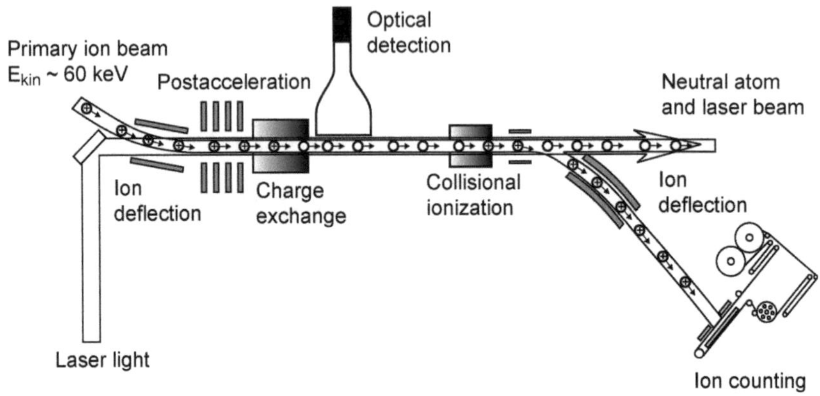

Figure 1. Experimental setup for collinear laser spectroscopy involving collisional ionization in a gas target and resonance detection of ions.

example: What is the influence of the proton magic number on the development of the radii around calcium; should one ascribe the difference between Ca and K to an odd–even effect in the proton number or is it due to an exceptional situation in the neighbourhood of the $Z = 20$ proton shell closure, and so on? In this context it is very interesting to extend the measurements of argon radii to the other neutron rich Ar nuclei in the $1f_{7/2}$ shell, in order to get a complete picture of the ms nuclear radii evolution in the calcium region.

The optical investigation of radii and moments in long chains of isotopes with $Z < 20$ is rather difficult because of the atomic properties of light elements and because of the low production yield of isotopes away from stability. For these reasons the experimental situation is very challenging. It requires high sensitivity and selectivity and very high spectroscopic accuracy. The experimental technique, used for the present study of the $f_{7/2}$ argon isotopes, has proven to meet these requirements.

2. Experiment

The experiments were performed at the ISOLDE on-line isotope separator at the CERN PS-Booster synchrotron. The isotope shift investigations were carried out by collinear fast-beam laser spectroscopy combined with an ultra-sensitive resonance detection technique. The method is based on neutralizing the ions in a charge exchange cell followed by optical pumping, state-selective collisional ionization and non-optical detection via ion or beta-decay counting [4]. The experimental setup is shown in Figure 1. The long lived argon isotopes between mass numbers $A = 40$ and 46 have been detected by direct ion counting. This technique provides a detection efficiency superior to the fluorescence photon counting and does not suffer from background due to the stray laser light.

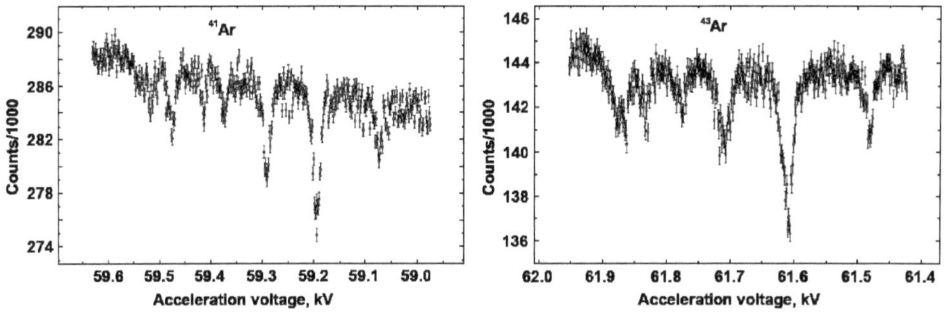

Figure 2. Hyperfine structure of the odd-*A* isotopes ^{41}Ar and ^{43}Ar.

Although its application is restricted to rather pure beams a mass resolution of about 1,500 was sufficient for the suppression of beam contaminants in the investigated argon region.

The atomic transition suitable for probing the nuclear structure of argon using this experimental technique is the transition $3p^54s\ [3/2]_2 \rightarrow 3p^54p\ [3/2]_2$ with λ = 763.7 nm [4]. Near resonant charge exchange between argon ions and potassium atoms in the charge exchange cell predominantly populates the metastable $3p^54s\ [3/2]_2$ state in atomic argon. As ionization of the metastable state is favoured in collisions with the gas target molecules, optical resonance pumping the atoms to the ground state lowers the ionization efficiency and the detected optical resonances appear as flop-out signals in the ion count rate. The involved efficiencies were typically 50% for the neutralization and 10% for the re-ionization with about 25% drop of the count rate in the single resonance of an even–even isotope.

In order to extract ms charge radii changes from isotope shifts in the argon system, the small nuclear field shift must be accurately determined. The two main experimental limitations, affecting the absolute accuracy, were found to be the laser frequency stabilization and the absolute determination of the total acceleration voltage. Special care was taken to keep the laser frequency stabilized, to prevent mode-hops and to eliminate long-term instabilities. The problem with the accuracy of the acceleration voltage was tackled by the development of a method for high-accuracy determination of the beam kinetic energy [5]. It is based on the excitation of two optical transitions, separated by twice the Doppler shift, at the same laser frequency in collinear and anti-collinear laser beam geometry, without changing the static part of the acceleration voltage. As both transitions are serving as secondary wavelength standard, their frequencies are very precisely known, and the kinetic energy of about 60 keV can be calculated to 0.3 eV from the resonances recorded within a narrow range of the Doppler-tuning voltage. This accuracy corresponds to systematic uncertainties of the isotope shift not exceeding 1 MHz for two neighbouring argon isotopes.

Examples of the recorded spectra are shown in Figure 2 for two odd-*A* argon isotopes. As can be seen, most hyperfine structure components are well resolved. The

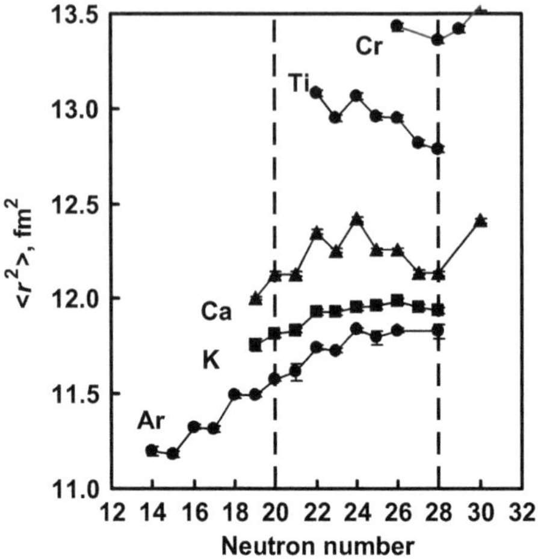

Figure 3. Mean square nuclear charge radii of argon, calcium, potassium, titanium and chromium nuclei versus neutron number.

resonances were modelled by a convolution of a Lorentzian profile with an exponential function which was found to describe most adequately the line shape [6].

3. Experimental results

Isotope shifts measurements, $\delta\nu^{38,A}$, were performed on the argon isotopes $^{38,40\text{–}44,46}$Ar and used to evaluate the mean square nuclear charge radii for the majority of the $f_{7/2}$ shell nuclei with $Z = 18$. Some special problems of extracting information about nuclear charge radii from isotope shifts of light elements are pointed in the previous paper on argon [4]. Because of these problems a deliberate treatment of the experimental data is necessary. The procedure of data analysis and evaluation will be described in detail in our forthcoming paper [7].

The results of the isotope shift measurements for the long sequence of Ar isotopes covering the upper νsd and nearly the whole $\nu f_{7/2}$ shell have been expressed in the corresponding nuclear radii and presented in Figure 3. The data on $^{41\text{–}44}$Ar have been obtained for the first time and should be regarded as preliminary. As can be seen, the charge radii evolution in the $f_{7/2}$ shell is characterised by (a) a non-symmetric parabola-like trend; (b) a 'normal' well pronounced odd–even staggering effect and (c) a slope of the curve steeper at the beginning of the interval between both neutron magic numbers and saturating at the end. These features are insensitive to the systematic uncertainties affecting only the overall scale of the radii changes [4, 7].

The experimental data are in a good qualitative agreement with the predictions of Klein *et al.* [4] based on the spherical HF approach with the Skyrme interaction. It describes explicitly the role of the configuration mixing within the proton sd shell, predicts qualitatively a steady increase of nuclear radii between both neutron shell closure $N = 20$ and $N = 28$, but fails to explain the parabolic shape and the odd–even staggering of the radii variation. In a more complete discussion the role of the collective effects has to be taken into account.

4. Nuclear charge radii in the calcium region – overview

The present data on Ar together with earlier data for K, Ca and the recent laser spectroscopy results on $^{44-50}$Ti [8] may serve as a cornerstone for a more global systematics of the nuclear radii in the calcium region. According to Figure 3 the well pronounced symmetric parabolic shape observed for the Ca case is not reproduced either in the lower-Z or in the higher-Z neighbouring isotope chains. Now it becomes more evident that the shape asymmetry for elements below and above Ca is in the opposite direction. As mentioned in the introduction, such behaviour is in contrast to what is expected from the nuclear radii systematics at N greater than 28 [9]. In general, as more experimental data become available, it is the behaviour of the Ca chain radii that should be considered as anomalous in this region. The strongly magic proton number, $Z = 20$, of Ca seems to reinforce the $N = 20$ and 28 closures, leading to reductions, particularly in the neutron, but also proton radii.

Very unusual is the behaviour of charge radii around the neutron shell closure $N = 20$ [4, 9]. The sequential addition of neutrons going from the sd- to the $f_{7/2}$-shell gives a smooth change of the successive charge radii, indicating the disappearance of any shell effect. This most striking feature has been explained by different authors [4, 10, 11] as arising from a cancellation of the monopole polarisation and the quadrupole polarisation of the proton core when valence neutrons are added.

The data in Figure 3 confirms the features of the odd-even staggering (OES) that were already discussed in Section 1. Away from Ca the OES decreases. The smallest OES is observed for the odd-Z isotopes of K. Whether this is a result of the pairing effect in the even–Z chains is a question that deserves closer attention than we are able to give now. For the even-Z neighbours of Ca the OES is smaller than in the calcium case. This may be due to the stabilising role of the closed proton shell of Ca.

The mean field calculations, which usually aim to describe nuclear parameters and radii over a large region of nuclear masses and charges, cannot account for the details of the isotope shifts in the calcium region. The predicted isotopic evolutions of the charge radii are usually featureless. However, there are approaches which are able to explain some of the observed features. To this category belong for example: (i) The calculations based on the HF method with Skyrme interactions [4] predicting the general trend of the Ar charge radii in the sp and fp shell; (ii) the shell model calculations of Caurier *et al.* [11] using cross-shell proton-neutron correlation being the best description of charge radii trend of Ca isotope; (iii) the self-consistent RMF approach [12, 13] predicting in the case of Ti a continual increase in the charge radius going from $N = 28$ to the $N = 20$ shell closure. The same RMF calculations do not predict a noticeable kink in the charge radii systematics at $N = 20$ for any isotope chain.

5. Conclusion

In conclusion we will briefly resume the main results of the present work. Isotopes far from stability have become accessible due to the application of the ultra-high-sensitive collinear fast-beam laser spectroscopy technique at the ISOLDE facility [4, 7]. The isotope shift measurements of $^{38,40-44,46}$Ar, performed in the present work, together with the data of Klein *et al.* [4], give insight on the development of the ms charge radii in a long sequence of argon isotopes. Thus, new information on the behaviour of the nuclear charge radius in a scarcely investigated region around the calcium ($Z = 20$) chain is obtained. The data may serve as a guide to incorporate essential new features into the theoretical approaches.

A fuller picture of the nuclear properties in the calcium region implies extending the experimental information. Here we note the importance of some further investigations:

A continuation of the optical investigation of Ti isotopes to the lighter ^{43}Ti and, especially, ^{42}Ti at the $N = 20$ shell is necessary to provide a stringent test of available theoretical models. Information on nuclear radii changes in the isotopic chain of Sc, the second neighbouring odd-Z element of Ca, will enforce a clearer picture of the role of the pairing effect of even-Z isotopes. These are upcoming experiments, planned to be carried out with the laser spectroscopy setup at the IGISOL facility.

An additional feature has to be emphasised.

Based on the data presented in Figure 3, isotonic nuclear radii systematics in the Ca-region could be evaluated [7]. However, it would be of great importance to extend the region of investigated elements to higher Z up to the proton magic number $Z = 28$ in order to obtain more detailed information about the isotonic behaviour of nuclear charge radii in the $1f_{7/2}$ proton shell. For these nuclei the proton drip line is accessible just in the $1f_{7/2}$ neutron shell. Due to the small binding energies of nuclei close to the drip lines, one expects a lowering of the nuclear density, novel type of shell structures, or presently unknown unusual collective modes.

On the contrary, investigations of the isotope chains covering the $\nu 1f_{7/2}$ shell for elements with $Z < 18$ will move us close to the neutron drip line of these elements. In particular, the role of the $N = 28$ gap in the exotic neutron rich isotopes with $Z = 14$, 16 and 18 is widely discussed in the theory [14]. In general, one predicts that the $N = 28$ shell closure persists, even if eroded by the large neutron excess.

Finally, a continuation of charge radii measurements to the neutron rich side of the isotopic sequences with $N > 28$ is also necessary to give a link to the charge radii development toward the next magic numbers $N = 50$.

Acknowledgements

This work was supported by the German Ministry for Education and Research (BMBF) under contract No. 06 MZ 962 I and by the European Union in the framework of the HPRI program. One of the authors (K.M.) expresses her gratitude to the Alexander von Humboldt Foundation for the financial support in her research visits in Mainz and CERN.

References

1. Otten E. W., *Treatise on Heavy Ion Science* **8** (1989), 517.
2. Palmer C. W. P., Baird P. E. G., Blundell S. A., Brandenberger J. R., Foot C. J., Stacey D. N. and Woodgate G. W., *J. Phys. B At. Mol. Phys.* **17** (1984), 2197.
3. Touchard F., Guimbal P., Büttgenbach S., Klapisch R., de Saint Simon M., Serre J. M., Thibault C., Duong H. T., Juncar P., Lieberman S., Pinard J. and Vialle J. L., *Phys. Lett. B* **108** (1982), 169.
4. Klein A., Brown B. A., Georg U., Keim M., Lievens P., Neugart R., Neuroth M., Silverams R. E., Vermeeren L. and the ISOLDE Collaboration, *Nucl. Phys. A* **607** (1996), 1.
5. Geithner W., Hilligsoe K. M., Kappertz S., Katko G., Keim M., Kloos S., Kotrotsios G., Lievens P., Marinova K., Neugart R., Vermeeren L., Wilbert S. and the ISOLDE Collaboration, *Hyperfine Interact.* **127** (2000), 117.
6. Mueller A. C., Buchinger F., Klempt W., Otten E. W., Neugart R., Ekström C., Heinemeier J. and the ISOLDE collaboration, *Nucl. Phys. A* **403** (1983), 243.
7. Blaum K., Geithner W., Lassen J., Lievens P., Marinova K., Neugart R. and the ISOLDE collaboration, Nuclear moments and charge radii of argon isotopes between neutron shell closures $N = 20$ and $N = 28$ (to be published).
8. Gangrsky Yu. P., Marinova K. P., Zemlyanoi S. G., Billowes J., Campbell P., Flanagan K. T., Forest D. H., Griffith J. A. R., Huikari J., Moore R., Nieminen A., Thayer H., Tungate G. and Äystö J., *J. Phys., G, Nucl. Part. Phys.* **30** (2004), 1089.
9. Nadjakov E. G., Marinova K. P. and Gangrsy Yu. P., *At. Data Nucl. Data Tables* **56** (1994), 133.
10. Barranco F. and Broglia R. A., *Phys. Lett., B* **151** (1985), 90.
11. Caurier E., Langanke K., Martínez-Pinedo G., Nowacki F. and Vogel P., *Phys. Lett. B* **522** (2001), 240.
12. Lalazisis G. A., Farhan A. R. and Sharma M. M., *Nucl. Phys. A* **628** (1998), 221.
13. Lalazisis G. A., Raman S. and Ring P., *At. Data Nucl. Data Tables* **71** (1999), 1.
14. Samanta C. and Adhikari S., *Phys. Rev. C* **65** (2002), 037301.

Hyperfine Interactions (2005) 162:109–114
DOI 10.1007/s10751-005-9224-y

Laser Spectroscopy and β-NMR Measurements of Short-Lived Mg Isotopes

MAGDALENA KOWALSKA FOR THE IS 427 COLLABORATION AT ISOLDE/CERN
Institut für Physik, Universität Mainz, D-55099 Mainz, Germany.

Abstract. The feasibility of studying the neutron-rich ^{29}Mg, ^{31}Mg, and ^{33}Mg isotopes has been demonstrated with the laser and β-NMR spectroscopy setup at ISOLDE/CERN. The values of the magnetic moment and the nuclear spin of ^{31}Mg are reported and reveal an intruder ground state. This proves the weakening of $N = 20$ shell gap and places this nucleus inside the so-called 'island of inversion.' The experimental setup and technique, as well as the results, are presented.

Key Words: electromagnetic moments, fine and hyperfine structure, polarised beams, properties of nuclei.

1. Introduction

Laser techniques can be very useful not only to investigate properties of atoms as a whole, but even to gain insight into the structure of atomic nuclei by using the hyperfine interaction between the electrons and the nucleus [1]. By probing the ground state properties of radioactive nuclei, such techniques can contribute considerably to our understanding of the nuclear structure. This is especially important in unique regions such as the 'island of inversion,' which covers neutron-rich nuclei around $N = 20$ and $Z = 12$, whose properties cannot be explained with the simple nuclear shell model, but indicate a weakening of the $N = 20$ shell gap between the nuclear *sd* and *fp* shells [2–5]. The odd-A neutron-rich radioactive Mg isotopes are relevant in this context, since with 12 protons and 17 to 21 neutrons, they lie inside or at the borders of this 'island.' They have been recently studied at ISOLDE/CERN by employing the techniques related to collinear laser spectroscopy [6, 7]. β-NMR measurements for ^{29,31}Mg on optically pumped ions reveal their nuclear g-factors, whereas a combination with measurements of the hyperfine structure allows an unambiguous determination of the ground state nuclear spin and parity. For ^{31}Mg, the measured spin-parity $I^{\pi} = 1/2^{+}$ implies an intruder ground state and proves the weakening of the $N = 20$ nuclear shell gap for Mg isotopes.

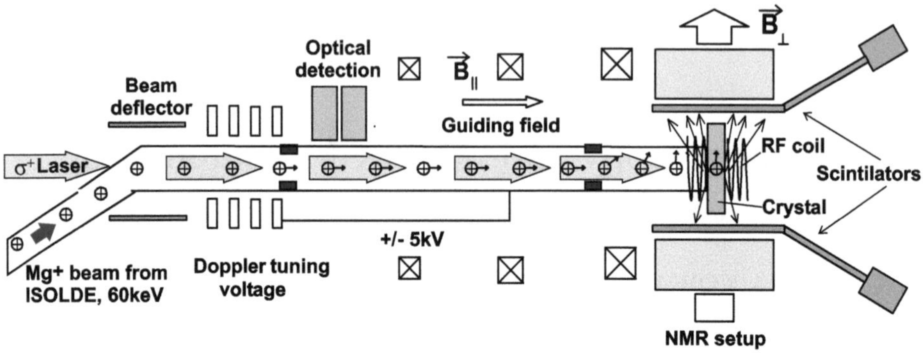

Figure 1. Experimental setup for laser and β-NMR spectroscopy on Mg ions.

2. Experimental method

Neutron-rich Mg isotopes are produced at the ISOLDE mass separator via fragmentation reactions in a uranium carbide target induced by a 1.4 GeV proton beam (typical intensity of 3×10^{13} protons per pulse, every 2.4 s). The produced atoms are then resonantly ionised with pulsed lasers (the RILIS method [8]), accelerated to 60 keV, and guided to the setup for the collinear laser spectroscopy, presented in Figure 1. The typical intensities at the experimental station are: 6×10^6, 2×10^5, and 9×10^3 ions/s of $^{29}Mg^+$, $^{31}Mg^+$, and $^{33}Mg^+$ (with respective half-lives of 1.3 s, 230 ms, and 90 ms). The above yields and half-lives make these Mg isotopes well suitable for β-asymmetry detection of optical pumping and for β-NMR experiments [4, 7].

The principle of the experimental method used for these short-lived isotopes is the following (Figure 1): The beam is polarised, implanted into a crystal and β decay electrons are detected. Because the nuclei are polarised, their decay is anisotropic and its asymmetry can be measured. In case of Mg beams, the polarisation is obtained via optical pumping [4, 9]. For this purpose they are overlapped in a weak external magnetic field with circularly polarised laser light, which causes transitions between the Zeeman levels of the hyperfine structure. For magnesium, optical polarisation is readily achieved for a singly charged ion in the excitation from the ground state ($3s\ ^2S_{1/2}$) to one of the first excited states, $3p\ ^2P_{1/2}$ or $3p\ ^2P_{3/2}$ (D_1 or D_2 line). The principle of the optical pumping is schematically shown in Figure 2 for $^{29}Mg^+$: In a weak external magnetic field positively polarised light (σ^+) induces atomic transitions with $\Delta m_F = +1$ (for σ^-, $\Delta m_F = -1$). Although the decay is isotropic and $\Delta m_F = -1, 0$ or $+1$, the laser light interacts many times with the same atom and causes excitations in which m_F can only increase by 1 (decreases by 1 for σ^-). As a result, most of Mg ions are in the ground state substate with the highest m_F (for σ^+, or lowest m_F for σ^-). A complication to this quite straightforward process is caused by the hyperfine pumping, in which the excited state decays to the other ground state

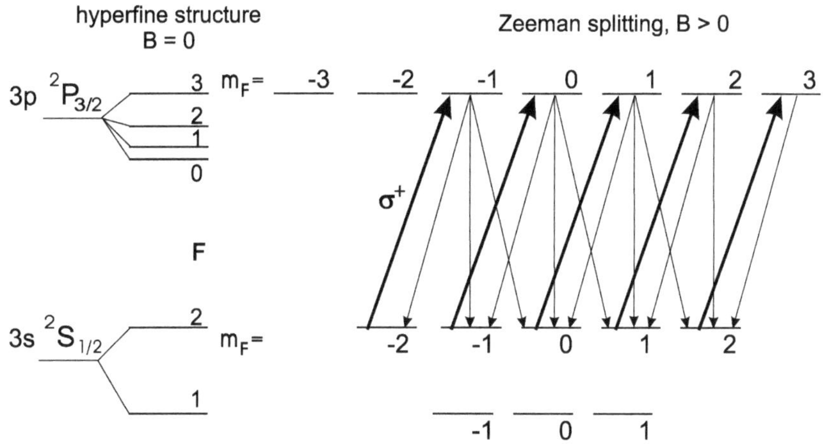

Figure 2. Optical pumping with σ^+ light for the D_2 line of $^{29}Mg^+$ ($I = 3/2$, $\mu_I > 0$).

level ($F = 1$ in Figure 2). This lowers the population of the ground state component available for optical pumping and therefore the polarisation is not equal to 100% [4].

The problem of efficient optical pumping is the laser power. The ultraviolet transition wavelength, 280 nm, was produced by the frequency doubling of visible dye laser radiation. The best doubling efficiency is obtained by an external resonator, therefore the experimental challenge is the setting and operation of a relatively complicated and delicate laser system for an online experiment. The fundamental beam is produced in a ring dye laser (Pyrromethene 556 as the active medium) pumped by a multiline Ar^+ laser. Powers up to 1 W at 560 nm were obtained with 6 W pumping. The output is then frequency doubled in an external *DeltaConcept* resonator with a BBO crystal and angle tuned phase-matching, and UV powers up to 15–20 mW are reached. Measurements with different laser powers show that the polarisation achieved by optical pumping is close to saturation (Figure 3).

Optical pumping creates longitudinal polarisation of the total (electronic and nuclear) spin system. Mg ions move further to a region of gradually increasing magnetic field, where the spins are adiabatically rotated and then decoupled, before entering a high transverse magnetic field (about 0.3 T) of an NMR-magnet. The ions are implanted into a crystal located in the middle of the magnet and their β decay is observed in a pair of plastic scintillators placed at 0° and 180° with respect to the magnetic field. Since β decay of a polarised ensemble is not isotropic, the amount of nuclear polarisation created via optical pumping is reflected in the experimental asymmetry, expressed as $A = (N_1 - N_2)/(N_1 + N_2)$, with N_1 and N_2 as β counts in the detectors.

One of the advantages of working with ions is the possibility that, instead of changing the laser frequency ν_L (requiring a relatively complicated calibration of the laser wavelength), one can modify the velocity of the ions. For this purpose

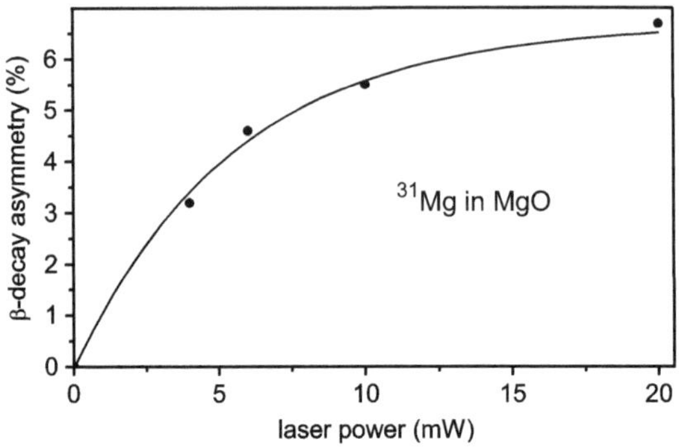

Figure 3. β-decay asymmetry as function of laser power. The polarisation achieved at 20 mW is close to saturation.

the optical pumping section forms a Faraday cage at a variable electrical potential U (Figure 1), which produces a Doppler shift of the laser frequency in the rest frame of the ions. The approach is known as 'Doppler tuning' [10] and the frequency ν in the frame of the ion is described by the following relation: $\nu = \nu_L(1-\beta)/\sqrt{(1-\beta^2)}$, where $\beta = v/c = \sqrt{1-(1+b)^{-2}}$ and $b = eU/mc^2$ (with the electric charge e, velocity of light c, and rest mass of the ion m). In the first step, this 'Doppler tuning' is used to record the hyperfine structure, which is observed in the change of the β asymmetry as a function of the acceleration voltage (Figure 4). From these measurements A and B hyperfine constants may be deduced, which in turn reveal the nuclear magnetic and quadrupole moments. (In case of the Mg ion, from the two lines used in this experiment, only the transition to the $^2P_{3/2}$ level can reveal the quadrupole moment, but because the hyperfine components of this state are not very well resolved, it is hard to obtain accurate information about the quadrupole moment in this way). After a hyperfine structure scan, the acceleration voltage is fixed at the hyperfine component giving largest asymmetry and NMR scans are performed: A radio-frequency field is produced by a coil placed around the host crystal, and when the frequency corresponds to the spacing ΔE_{mag} between different m_I substates of the studied nucleus, transitions between these levels occur. The redistribution of the m_I substates reduces the nuclear polarisation, seen in the disappearance of the experimental asymmetry. If a cubic host crystal is chosen, the resonance radio-frequency corresponds to the Larmor frequency, given by $\Delta E_{mag} = \nu_L h = g_I \mu_N B$ (with the nuclear g-factor g_I, nuclear magneton μ_N and external magnetic field B). For non-cubic host crystals, the resonance at the Larmor frequency is symmetrically split in $2I$ components by the interaction of the quadrupole moment Q_S with the internal electric field gradient V_{zz}, with the splitting proportional to $Q_S V_{zz}$. In this way, very accurate measurements of the nuclear magnetic and quadrupole moments can be performed.

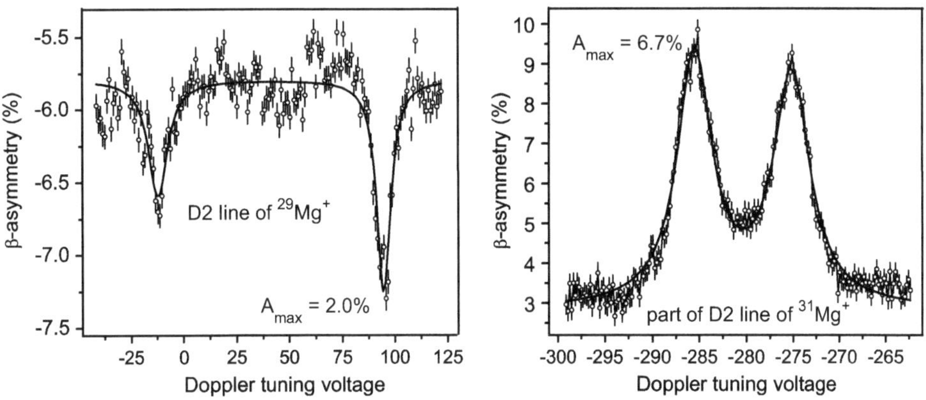

Figure 4. The experimental β-asymmetries in the D_2 line of $^{29}Mg^+$ and $^{31}Mg^+$ as a function of Doppler tuning voltage.

3. Experimental results

The first measurements on Mg beams were performed to test the optical polarisation for $^{29}Mg^+$ and $^{31}Mg^+$. For this purpose different cubic host crystals were used. MgO gave the highest asymmetries at room temperature (up to 6.7% for ^{31}Mg in the D_2 line) and was used in the experiments. The other tested crystals were Pt and Au, giving 3.1 and 1.8% asymmetry (also for ^{31}Mg), respectively. These tests showed that optical pumping experiments with Mg beams provided by ISOLDE are feasible (Figure 4).

After hyperfine scans, NMR measurements in MgO followed for both ^{29}Mg and ^{31}Mg. At the time of the workshop (May 2004) their analysis was still in progress, therefore no results were presented. This analysis has been now completed for ^{31}Mg and is published [11]. A brief summary of the results is presented below.

The highlights of these measurements are the ground state properties of ^{31}Mg, especially the unknown nuclear spin. By β-NMR we determined the nuclear g-factor. Calibration of the magnetic field was performed on ^8Li ($g(^8Li) = 0.826780(9)$ [12]) available from the same ISOLDE target, and it yields $\nu_L(^8Li) = 1807.03(2)$ kHz. Measurements on ^{31}Mg give $\nu_L(^{31}Mg) = 3859.73(18)$ kHz (Figure 5). The resulting value of the g-factor of ^{31}Mg (corrected for diamagnetism) is thus $|g| = 1.7671(3)$ [11]. The final error includes a systematic uncertainty accounting for the inhomogeneities of the magnetic field and its drift between the measurements on ^{31}Mg and ^8Li.

The magnitude of the hyperfine splitting is proportional to the product of g and I (and the position of particular components depends on the sign of g). By combining the NMR and hyperfine results we are able to determine the spin and parity of ^{31}Mg ground state to be $1/2^+$ [11]. This result requires at least two neutrons in the *fp* shell, which shows the intruder character of the ground state and proves the weakening of the $N = 20$ shell gap.

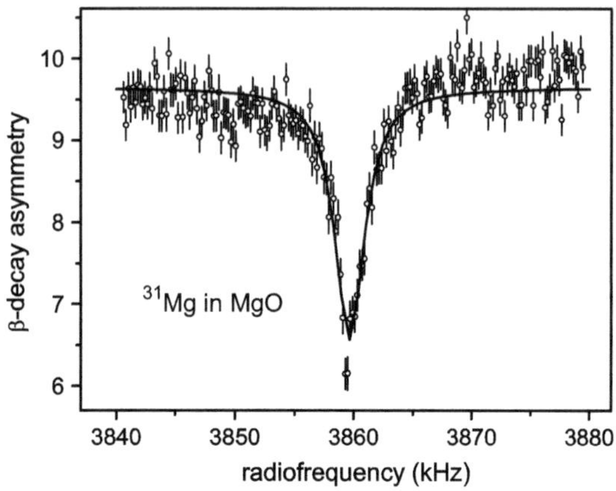

Figure 5. Typical β-NMR resonance of ^{31}Mg implanted in MgO.

Future plans include measurements of the magnetic and quadrupole moment of ^{33}Mg, as well as isotope shifts between different neutron-rich isotopes of magnesium.

Acknowledgments

This work has been supported by the German Ministry for Education and Research (BMBF) under contract No. 06MZ175, by the IUAP project No. p5-07 of OSCT Belgium, by the FWO-Vlaanderen, by Grant-in-Aid for Specially Promoted Research (13002001) and the EU under EURONS project (no. 506065). The authors thank the ISOLDE technical group for their assistance during the experiment.

References

1. Neugart R., *Eur. Phys. J. A* **15** (2002), 35.
2. Thibault C. *et al.*, *Phys. Rev. C* **12** (1975), 644.
3. Detraz C. *et al.*, *Phys. Rev. C* **19** (1979), 164.
4. Keim M. *et al.*, *Eur. Phys. J. A* **8** (2000), 31.
5. Guillemaud-Mueller D. *et al.*, *Nucl. Phys. A* **426** (1984), 37.
6. Neugart R., *Nucl. Inst. Meth.* **186** (1981), 165.
7. Geithner W. *et al.*, *Phys. Rev. Lett.* **83** (1999), 3792.
8. Köster U. *et al.*, *Nucl. Instr. Meth. B* **204** (2003), 347.
9. Arnold E. *et al.*, *Phys. Lett. B* **197** (1987), 311.
10. Kaufman S. L., *Opt. Comm.* **17** (1976), 309.
11. Neyens G. *et al.*, *Phys. Rev. Lett.* **94** (2005), 22501.
12. Raghavan P., *At. Data Nucl. Data Tables* **42** (1989), 189.

Hyperfine Interactions (2005) 162:115–123
DOI 10.1007/s10751-005-9215-z

Pumping ^{229m}Th by Hollow-Cathode Discharge

T. T. INAMURA[1,*], T. MITSUGASHIRA[2],
and THE OARAI COLLABORATION

[1]*Accelerator Research Facility, RIKEN, Wako-shi, Saitama 351-0198, Japan;
e-mail: tinamura@riken.jp*
[2]*Oarai branch, Institute for Materials Research, Tohoku University, Oarai-machi, Ibaraki 311-1313, Japan*

Abstract. An extremely low-lying nuclear isomer in ^{229}Th is still to be established in spite of efforts made since the first postulation of the isomer in 1976. To study such an extremely low-lying nuclear isomer, we have built an entirely new device that is based on the well-established atomic spectroscopic technique: Hollow-cathode discharge. The isomer in question is expected to be populated by the process NEET through atomic excited states if it exists in an energy range of electron volt. After switching off the discharge, we will measure alphas from the isomer, photons from the isomer itself and photons from atomic states to which the isomer is expected to decay by the inverse NEET. Our experiment is under way at Oarai using the world's radiochemically purest ^{229}Th sample.

Key Words: alphas and photons, NEET by electric discharge.

1. Introduction

It was in 1976 that Kroger and Reich postulated the 1st excited state in ^{229}Th at an energy < 100 eV [1]. Eighteen years later Helmer and Reich revised that excitation energy to 3.5 ± 1.0 eV [2]. Meanwhile, Burke et al. suggested the existence of such a low-energy isomer at ≤ 5 eV by observing angular distributions of the ^{230}Th(d,t)^{229}Th reaction [3]. Their results are, however, not direct evidence to support the existence of such an extremely low-lying isomer ^{229m}Th. Several attempts have been made to obtain direct evidence for the isomer ^{229m}Th. Because of its excitation energy comparable to the UV, mostly optical measurement has been carried out using ^{233}U that decays to ^{229}Th by emission of alphas [4–7]. None of them are successful. Recently, two different types of attempts have been made to determine the lifetime of ^{229m}Th by avoiding the huge background inevitably caused by the parent nucleus ^{233}U [8, 9]. Unfortunately, their measurements are inconsistent with each other, and neither

* Author for correspondence.

of them has produced clear evidence for the existence of ^{229m}Th. Last year Barci et al. reported nuclear structure of ^{229}Th from extensive γ-ray spectroscopy of ^{233}U α-decay with conventional LEPS and HPGe counters, suggesting that their measurements are consistent with the existence of the 3.5-eV isomer [10]. However, that is not direct evidence for the isomer in question, either.

Here we shall present a unique combination of a novel idea and nuclear chemists' expertise, aiming at unprecedented achievements in studying the exotic isomer ^{229m}Th.

2. How to do pumping ^{229m}Th

As is seen from Figure 1, the novel idea to populate ^{229m}Th was forged by thinking of the underlying interactions in Coulomb excitation of atomic nuclei, hyperfine interaction including isotope shifts, and the atomic excitation mechanism in the hollow-cathode discharge tube (lamp). The underlying interactions are electromagnetic interactions that can basically be described in the same form of Hamiltonian as, to first order in v/c,

$$H_{1,2} = \iint\limits_{\tau_1\tau_2} \frac{\rho_1(\mathbf{r}_1)\rho_2(\mathbf{r}_2)}{|\mathbf{r}_1 - \mathbf{r}_2|} d\tau_1 d\tau_2 - \frac{1}{c^2} \iint\limits_{\tau_1\tau_2} \frac{\mathbf{j}_1(\mathbf{r}_1)\mathbf{j}_2(\mathbf{r}_2)}{|\mathbf{r}_1 - \mathbf{r}_2|} d\tau_1 d\tau_2 \tag{1}$$

where $\rho(\mathbf{r})$ and $\mathbf{j}(\mathbf{r})$ are charge density and current density of interacting entities denoted by subscripts 1 and 2, respectively. Note that we are dealing with excitation energies of the order of electron volt.

In Figure 1c a broad belt indicates that there are to a great extent multiple collisions between the incoming ions I (here singly charged Ar ions) and the sample atoms $_Z$A (here ^{229}Th): The Ar gas pressure will be 100–200 Pa, the discharge voltage 200–300 V, and the current about 100 mA. This discharge condition in the hollow-cathode tube is very much like a standard and one can excite atomic states up to about 10 eV in A I (neutral atom corresponding to the first term in Figure 1c) and A II (singly charged atom corresponding to the second term in Figure 1c). We confirmed that by a preliminary experiment with natural Th using the UV spectrograph at Poznań University of Technology in 2002 [11] (see Appendix A). Eventually, one can expect one of atomic states thus excited will be at near resonance with a nuclear isomer if it exists. This is analogous to the hyperfine interaction (Figure 1b) between atomic states and a long-lived nuclear state in terms of electromagnetic interactions. More correctly, this process causing a kind of resonance is the inverse internal conversion, NEET [12]. Here is no electron emission, so that it may be called extended NEET as Karpeshin *et al.* suggested [13, 14]. Once such an atomic state is populated, Karpeshin's formulae can be applied to estimate the population of the isomer in question, i.e., ^{229m}Th.

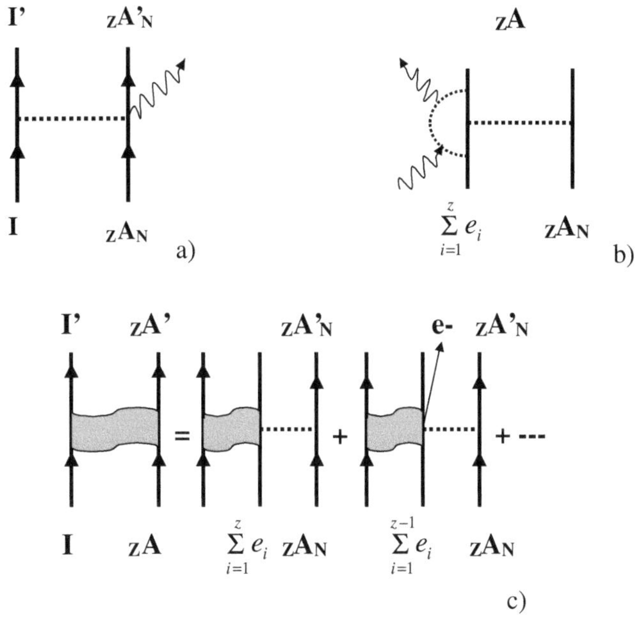

Figure 1. Feynman diagrams: (a) Coulomb excitation, (b) hyperfine interaction including isotope shifts, and (c) pumping mechanism of a nuclear isomer in the energy range of eV. $_{Z}A'$ and $_{Z}A'_{N}$ indicate excited states for the atom $_{Z}A$ and the nucleus $_{Z}A_{N}$, respectively. *Broad belts* in (c) indicate that there is to a great extent a multiple scattering process. All interactions are purely electromagnetic.

The A I and A II are such systems as

$$\left(\sum_{i=1}^{z} e_i\right) \otimes {}_{Z}A_N \quad \text{and} \quad \left(\sum_{i=1}^{z-1} e_i\right) \otimes {}_{Z}A_N, \tag{2}$$

respectively. For simplicity, let us take the ground-state atomic electron configuration of A II, i.e., Th^{+1}: $7s^2 6d_{3/2}$. And there is an almost degenerated state $7s(6d_{3/2})^2$ at an excitation energy of 0.031 eV. Based on these atomic information, as mentioned above, we can expect atomic excited states within 10 eV from which NEET processes will populate the isomer $_{Z}A'_{N}$ in the nucleus $_{Z}A_{N}$ (Th). According to Karpeshin *et al.* [13], the $8p_{1/2}$ state at about 7 eV in A II is likely to cause the NEET in case of the isomer in an energy range of 2–4 eV. If the isomeric excitation energy is 3.5 eV as postulated by Helmer and Reich [2] and the nuclear transition is of M1 as expected from Nilsson orbital assignments to the ground state and the isomer, $5/2^+$ [633] and $3/2^+$ [631], then we have the NEET probability $P_{\text{NEET}} \sim 1.4 \times 10^{-10}$ (see [13] for details).

From the preliminary experiment [11], the number of atoms at the $8p_{1/2}$ state will be of the order of 10^{13}. (Here parameters are: For a plate WU1 with sensitivity of about 24 DIN (200 ASA), or $\geq 10^6$ photons/line; the exposure time

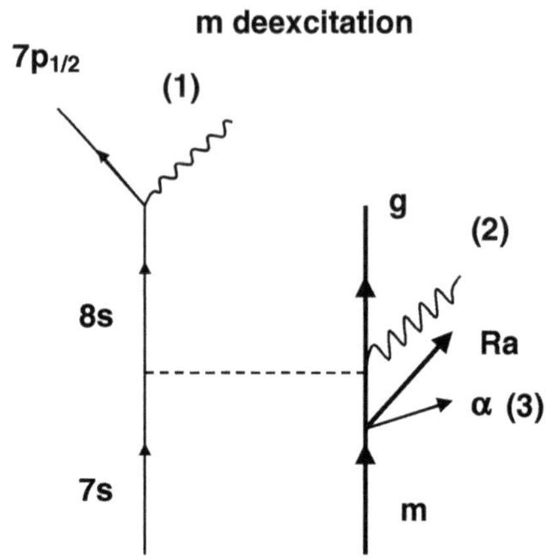

Figure 2. Deexcitation modes of ^{229m}Th: (1) indirect emission of photons; (2) direct emission of photons; and (3) α decay.

is 3 min; the effective solid angle for the UV spectrograph is about 10^{-9}). Let us take the half-life of 14 h [9] for the isomer, and then the isomer population N_i at 10-h discharge will be 3×10^7 atoms.

The great advantage in the present method is that there is no need to know exactly the excitation energy of ^{229m}Th. In other words, a laser spectroscopic technique is applicable to populate such an isomer in principle, but practically it is impossible without knowing the exact excitation energy.

3. What to measure

The isomer m in question can decay by three modes (see Figure 2): 1) Indirect emission of photons (internal conversion process without emitting atomic electrons); 2) direct emission of photons (M1 nuclear transition); and 3) α decay. In case of mode 1 for Th II the initial state is given by a coupled state of the atomic |7s> and the isomer |m>, i.e., |7s>|m> and the final state |8s>|g> where |g> is the nuclear GND. Modes 2 and 3 can be treated independently of the atomic states.

Mode 1 is TEEN according to Karpeshin [15] and this process is expected to be dominant [16] so that one will be able to detect delayed photons through atomic deexcitation, for example, |$7p_{1/2}$> to the uncoupled atomic GND |$6d_{3/2}$> or its degenerate counterpart |7s> as well as mode 2. However, first we will measure α decay (mode 3) to avoid photon background. When it comes to the S/N ratio, α measurement is best and it is easy to detect even a few alphas per hour practically free from background.

From the α-decay systematics in this mass region, one can expect four dominant α decays from ^{229m}Th to ^{225}Ra (mode 3) [9]. The α-decay half-life is thus estimated to be about 500 years for ^{229m}Th, i.e., the partial decay constant λ_α

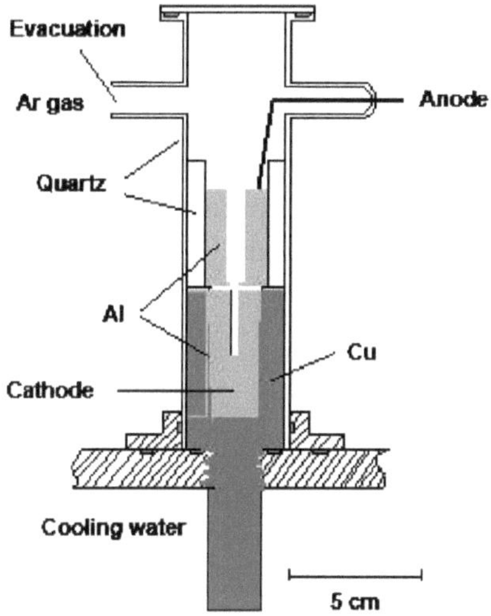

Figure 3. Schematic cross-sectional view of our hollow-cathode discharge tube.

$\sim 4.39 \times 10^{-11}$ s^{-1}. (It is known that the ground state of ^{229}Th has $T_{1/2}$ = 7,340 years, i.e., $\lambda_\alpha = 2.99 \times 10^{-12}$ s^{-1}). Now we can have a counting rate of alphas from the isomer, $\lambda_\alpha N_i \sim 1.3 \times 10^{-3}$ s^{-1}, i.e., 4 ~ 5 counts/h. This is a rather pessimistic estimation because we did not include a possible enhancement factor for P_{NEET} [13]. This NEET process is repeated during the isomer lifetime and the repetition period will be of the order of 0.1 μs. Consequently, one can expect such an enhancement factor as 5×10^{11} for $T_{1/2}$ = 14 h. However, only the experiment will tell us the truth. (Note: One cannot have the counting rate increased by the enhancement factor straightforwardly. There is a limit by the saturation of population of the isomer).

Energies of the four dominant alphas from ^{229m}Th should be 4930 keV (transition to the $J^\pi = 3/2^+$ ($3/2^+$ [631]) level of ^{225}Ra at 149.96 keV), 5,036 keV (to $J^\pi = 3/2^+$ ($1/2^+$ [631]) at 42.77 keV), 5,053 keV (to $J^\pi = 5/2^+$ ($1/2^+$ [631]) at 25.41 keV), and 5,079 keV (to $J^\pi = 1/2^+$ ($1/2^+$[631]) Ra GND) [9]. The dominant α component (56%) of the ^{229}Th α decay has an energy of 4,845.3 ± 1.2 keV [9] and other significant components have energies definitely lower than 5,000 keV. It should also be stated that the energies of alphas from the daughters are higher than 5,730 keV, which do not interfere with measurement of alphas of interest. Accordingly, we can distinguish the alphas from ^{229m}Th and those from ^{229}Th. When the lifetimes of these states are taken into account, it should be made much easier than with energy information only.

To measure alphas emitted by mode 3, we built a hollow-cathode discharge tube as shown in Figure 3 (see Appendix B for details). A sample of ^{229}Th is placed inside the hollow cathode (the lower part of Al electrodes). Its amount is about 0.1 μg or 1 kBq. Special care was exercised to quickly make the measurement after switching off the discharge (Appendix C). It is possible to start the measurement within 5 min. If we do chemical treatment to have a better α source of ^{229}Th/^{229m}Th, it will take about 15 min (Appendix D).

The experiment is under way at the Oarai branch, IMR, Tohoku University using the world's radiochemically purest ^{229}Th (Appendix E). We are also planning to measure photons (Appendix C).

Acknowledgements

The author (T.I.) would like to acknowledge Dubna–Poznań–St. Petersburg–Warsaw collaboration members of which first supported this idea: Yu. Gangrsky, B. Markov, S. Zemlyanoi (Flerov Lab., Dubna), K. Gromov (LNP, Dubna), W. Chmielowski (Poznań Univ., Dubna); B. Archmowicz, J. Dembczynski, E. Stachowska (Poznań Univ. Tech.), Z. Błaszczak (Poznań Univ.); F. Karpeshin (St. Petersburg Univ.), M. B. Trzhaskovskaya (NPI, Gatchina); and S. Chojnacki (HIL, Warsaw Univ.), J. Żylicz (IEP, Warsaw Univ.). When he visited Poznań University of Technology about five years ago, the author saw a hollow-cathode lamp coupled with a UV spectrograph and this idea came to him. Eventually he organized that collaboration (see Appendix A).

Thanks are also due to the members of the Oarai collaboration: T. Ohtsuki, H. Yuki (Nuclear Science lab., Tohoku Univ., Sendai); M. Hara (Oarai branch, IMR, Tohoku Univ.); H. Haba (RIKEN), H. Kikunaga, T. Nakanishi, A. Yokoyama (Kanazawa Univ.); K. Takamiya (Research Reactor Inst., Kyoto Univ.); and Y. Kasamatsu, A. Shinohara (Osaka Univ.).

Appendix A. Preparatory work of Dubna–Poznań–St. Petersburg–Warsaw collaboration

As the initiation of the Dubna–Poznań–St. Petersburg–Warsaw collaboration, a preparatory measurement was carried out with natural Th, i.e., ^{232}Th, using a UV spectrograph *Carl Zeiss 'Jena'* at Poznań University of Technology in 2002 [11]. The purpose of this measurement was to test the feasibility of detecting delayed photons from ^{229m}Th in question in the range of 3.5 ± 1.0 eV [2] (280–500 nm) by confirming the excitation of atomic states $8p_{1/2}(6d_{3/2})^2$ and $8p_{3/2}(6d_{3/2})^2$ at around 7 eV in Th II by the hollow-cathode discharge, which are considered as prerequisite for the NEET to populate ^{229m}Th [13].

A hollow-cathode discharge lamp for the UV spectrograph was loaded with a natural Th foil (25 mm h × 20 mm w × 0.025 mm d). The aluminum hollow cathode was water-cooled. The depth of the hollow was 25 mm, its diameter 6.5 mm, and the 20-mm wide Th foil was rolled and placed inside the hollow to line neatly. The discharge condition was as follows: Ar gas at about 100 Pa, V = 160 V and A = 70 mA.

Photographic plates WU1 were used to detect photons: The plate WU1 is guaranteed sensitive to photons with wavelengths from 250 to 450 nm (2.8–5.0 eV); its sensitivity is about 24 DIN (200 ASA), or $\geq 10^6$ photons/line. The exposure time was 3 min. Measurements were made in such a way that first, the 250–330-nm range was recorded, second, the 310–390-nm range, and third, the 370–450-nm range, by adjusting the grating position.

A great many fluorescence spectral lines of Th I and Th II were observed in each range. By comparing some of prominent lines with transitions between computed atomic states in Th II and those given in [13], it was concluded that all atomic states up to about 10 eV were populated, some of which are likely to be the states $8p_{1/2}(6d_{3/2})^2$ and $8p_{3/2}(6d_{3/2})^2$. The similar expectation should be equally valid for Th I although transitions are more complicated. It should also be noted that fluorescences from Th I as well as Th II were observed in the range from about 310 to 1100 nm using a hollow-cathode lamp [17].

Appendix B. Hollow-cathode discharge tube

When designing the hollow-cathode tube, we exercised special care to handle radioactive ^{229}Th and ^{229m}Th. A schematic cross-sectional view of the tube is shown in Figure 3. There is an empirical golden rule in designing to produce a well-localized plasma in the hollow (B. Arcimowicz at Poznań University of Technology, private communication): The ratio of the depth of the hollow l to its diameter ϕ, $l/\phi \geq 4$ and the gap d between the anode surface and the cathode surface should be considerably smaller than ϕ. Our present measurements are: l = 25 mm, ϕ = 5 mm, and d = 1 mm. The cathode is water-cooled. Cooling is effective not only for safety reason but also for localizing the plasma in the hollow.

The hollow-cathode, the anode, and other parts in the tube are quickly dismantled to take out the radioactive ^{229}Th sample in the hollow after switching off the discharge.

Appendix C. Preparatory work for alpha and photon counting

ALPHA COUNTING

A simple and quickest way will be the use of an aluminum foil or rhenium foil, which is rolled to line inside of the hollow cathode and cut in half. Before inserting the foil cut in half, its one piece (or both pieces) is deposited on with about 0.1 μg ^{229}Th: A drop of 10 μl of 8 M HNO_3 solution of ^{229}Th whose concentration is 10 ng/μl is dried on the foil piece.

After being subjected to the discharge in the hollow-cathode tube for a certain period (1 to 10 h), the foil piece will be taken out from the hollow-cathode tube to be placed in an α counting vacuum chamber. Usually we use a Si detector with 450-mm^2 active area and 100-μm depleted depth. Its energy resolution is 17 keV, full width at half maximum (FWHM) for ^{241}Am 5486-keV alphas in vacuum at room temperature.

PHOTON COUNTING

We have a setup for photon counting that consists of a concave mirror and a photomultiplier Hamamatsu R585 that has good spectral response in the range from 160 to 650 nm.

The foil piece described above will be placed in the setup. First, it will be placed just in front of the photomultiplier for the measurement as close as possible.

Appendix D. How to make a better α source of ^{229}Th/^{229m}Th

If the energy resolution in the quick α counting is not good enough for our purpose, we will prepare a better Th α source by means of the following chemical procedure [18].

The Th deposited on the foil or directly inside the hollow cathode without using the foil is dissolved with 8 M HNO_3, and is purified by the anion-exchange method using a chromatographic column (5 mm $\phi \times 40$ mm) filled with the anion-exchange resin (Dowex 1X8, 200–400 mesh): First, the Th in 8 M HNO_3 is trapped by the resin; second, the resin is washed with 25 ml of 8 M HNO_3 to remove impurities; and third, the Th thus purified is washed out with 10 ml of 1 M HCl. The Th in the eluate is co-precipitated with a small amount of Sm as hydroxide with an excess amount of aqueous NH_3. The precipitate filtered on a filter paper (Whatman, Anodisc25) 18 mm in diameter is sintered on a hot plate to provide a better α source of ^{229}Th/^{229m}Th. The energy resolution thus obtained is found to be about 22-keV FWHM at 8376-keV alphas from the daughter nucleus ^{213}Po.

It takes about 15 min to complete the whole chemical procedure.

Appendix E. ^{229}Th Source

Our ^{229}Th source was obtained from the mother ^{233}U sample in a HNO_3 solution by anion-exchange chromatography in May 1980. The isotopic impurity ^{232}U in the mother sample was known to be 3 ppm by alpha spectroscopy. This impurity caused contamination of the separated ^{229}Th by ^{228}Th ($T_{1/2}$ = 1.913 year) at the beginning: In terms of alpha activity ^{228}Th was about 15 times as much as ^{229}Th. Today, however, ^{228}Th has effectively decayed away to such an extent that it is undetectable. Radiochemically speaking, our ^{229}Th source is almost 100% pure. The ^{229}Th source subjected to the discharge in the hollow-cathode tube could further be purified by anion-exchange chromatography (see Appendix D).

References

1. Kroger L. A. and Reich C. W., *Nucl. Phys. A* **259** (1976), 29.
2. Helmer R. G. and Reich C. W., *Phys. Rev. C* **49** (1994), 1845.
3. Burke R. G., Garret P. E., Tao Qu and Naumann R. A., *Phys. Rev. C* **42** (1990), R499.
4. Irwin G. M. and Kim K. H., *Phys, Rev. Lett.* **79** (1997), 990.
5. Richardson D. S., Denton D. M., Evans D. E., Griffith J. A. R. and Tungate G., *Phys. Rev. Lett.* **80** (1998), 3206.
6. Utter S. B., Beiersdorfer P., Barnes A., Loughheed R. W., Crepso López-Urrutia J. R., Becker J. A. and Weiss M. S., *Phys. Rev. Lett.* **82** (1999), 505
7. Shaw R. W., Young J. P., Cooper S. P. and Webb O. F., *Phys. Rev. Lett.* **82** (1999), 1109.
8. Browne E., Norman E. B., Canaan R. D., Glasgow D. C., Keller J. M. and Young J. P., *Phys. Rev. C* **62** 014311, (2001).
9. Mitsugashira T., Hara H., Ohtsuki T., Yuki H., Takamiya K., Kasamatsu Y., Shinohara A., Kikunaga H., and Nakanishi, T. J., *Radioanal. Nucl. Chem.* **255** (2003), 63.
10. Barci V., Ardisson G., Brarci-Funel G., Weiss B., El Samad O. and Sheline R.K., *Phys. Rev. C* **68** 034329, (2003).
11. Arcimowicz B., Chojnacki S. and Inamura T. T., Warsaw University Heavy Ion Laboratory Annual Report 2002 (2003) p. 43.
12. Morita M., *Prog. Theor. Phys.* **49** (1973), 1574.
13. Karpeshin F. F., Band I. M., Trzhaskovskaya M. B. and Listengarten M. A., *Phys. Lett. B* **372** (1996), 1.
14. Inamura T. T., Karpeshin F. F. and Trzhaskovskaya M. B., *Czech. J. Phys.* **53** (2003), B349.
15. Karpeshin F. F., Band I. M. and Trzhaskovskaya M. B., *Nucl. Phys. A* **654** (1999), 579.
16. Karpeshin F. F., Band I. M., Trzhaskovskaya M. B. and Pastor A., *Phys. Rev. Lett.* **83** (1999), 1072.
17. Valero F. P. J., *J. Opt. Soc. Am.* **58** (1968), 484; **58** (1968), 1048.
18. H. Kikunaga, Doctoral thesis at Kanazawa University, to be published.

Hyperfine Interactions (2005) 162:125–132
DOI 10.1007/s10751-005-9213-1

Resonance Conversion as the Predominant Decay Mode of ^{229m}Th

F. F. KARPESHIN[1,*] and M. B. TRZHASKOVSKAYA[2]
[1]*Fock Institute of Physics, St. Petersburg State University 198504, St. Petersburg, Russia*
e-mail: Feodor.Karpeshin@pobox.spbu.ru; karpesh@nuclpc1.phys.spbu.ru; Feodor.Karpeshin@na.infn.it
[2]*Petersburg Nuclear Physics Institute, Gatchina, Leningrad district 188300, St. Petersburg, Russia*

Abstract. Calculations performed within the advanced MCDF method show that the probability of decay of the 3.5-eV level of ^{229m}Th through the resonance electronic bridges exceeds the direct radiative nuclear decay width by a factor of around $5 \cdot 10^3$. Experimental consequences for detection of the delayed soft photons or α particles from the decay of the isomer are discussed.

Key Words: 3.5-eV level of ^{229m}Th, electronic bridges, NEET, nuclear isomers, resonance internal conversion.

1. Introduction

Internal conversion (IC) strongly affects the nuclear lifetimes in excited states. In the case of conversion, the atomic electrons receive the nuclear transition energy in a nonradiative II kind process, and leave the atom. Internal conversion is an independent decay mode. Therefore, it diminishes the nuclear lifetime, accelerating the decay. Sometimes, the effect can be very strong.

On the other hand, IC is a reliable method of investigation of the nuclear structure, spins and parities of nuclear states. In practical studies, internal conversion coefficients (ICC) are usually exploited. They are defined as the ratio of the conversion Γ_c and radiative Γ_γ nuclear widths:

$$\alpha^{(\tau L)} = \frac{\Gamma_c(\tau L)}{\Gamma_\gamma(\tau L)}\,, \tag{1}$$

τL stand for the type and multipole order of the transition.

Usually ICC only weakly depend on the nuclear structure. They, however, reach high values, typically of 7–8 orders of magnitude, especially in the case of the high-multipole ($L = 3,4$) low-energy transitions. Thus, in the case of the 76-eV transition in ^{235}U, IC decreases the nuclear lifetime by 19 orders of

* Author for correspondence.

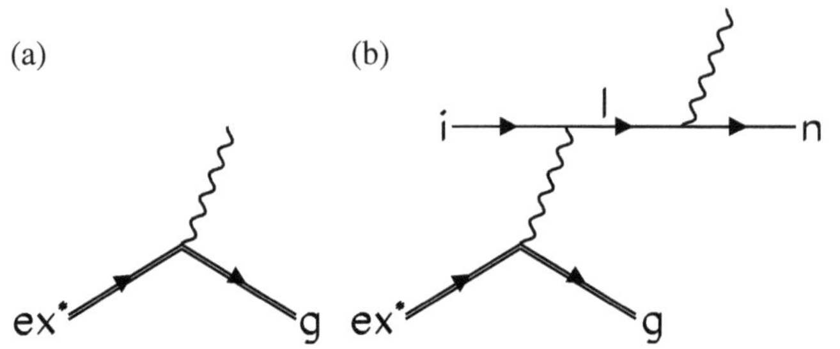

Figure 1. Feynman graphs of the radiative (a) and resonance conversion (b) decay of the 3.5-eV isomer of ^{229m}Th.

magnitude. For this reason, IC plays important part in the decay of isomers, just because their nature is often based on the high multipole order or low energy of the transition.

The uniquely low 3.5-eV isomer of ^{229}Th is of special interest among the nuclear isomers. In spite of many efforts, nobody could surely detect its decay. We note, however, that many attempts were based on a rather long estimated lifetime of the isomer, within 10 h, aimed at search for soft nuclear photons with the energy of $\sim$ 3.5 eV. It was however shown in [1, 2] that the main decay mode occurs via the resonance IC (Figure 1). In this process, the nuclear energy is also transferred to the atomic electrons. The energy is, however, too small for the electrons to leave the atom. Instead, one or more photons are reemitted, bringing away the energy of the atom. In the case of ^{229}Th, the most probable is the $7s - 8s - 7p - 7s$ chain of the electron transitions [1], with the two photons emitted with the energy crudely half the nuclear energy each. In this case, subthreshold, or discrete, bound, or resonance conversion takes place [3]. Resonance conversion factor is defined which plays the same role in the case of bound conversion, as usual ICC in the case of traditional conversion. This reads

$$R(J_i \to J_i^{(\ell)}) = \frac{1}{2\pi} \frac{\alpha_d^{M1}(J_i \to J_i^{(\ell)})\Gamma^{(\ell)}}{(\omega_n - \epsilon_\ell)^2 + (\Gamma^{(\ell)}/2)^2}, \tag{2}$$

where ω_n is the nuclear transition energy, ϵ_ℓ is the energy of the level ℓ with respect to the ground level of the Th atom, and $\Gamma^{(\ell)}$ is the radiative width of the ℓ–th level. In Equation (2), α_d^{M1} is defined by means of the same formulae as the conventional ICC (Equation (1)), with the only replacement of the wave function of the converted electron in the continuum by that in a discrete state. Consequently, α_d^{M1} acquires the dimension of energy, and cannot serve as ICC anymore. However, multiplied by the Breit–Wigner factor, in the form of R in Equation (1) it regains its physical sense of the ratio of the conversion and radiative probabilities ([4, 5]).

The conversion factor (2) was shown to be within one to three orders of magnitude, depending on the resonance defect $\Delta = \omega_n - \epsilon_l$ (Figure 1). This effect called the resonance electronic bridge decreases the expected lifetime to an amount of one minute or less [6].

In the present paper we investigate the resonance conversion mechanism within the framework of the advanced atomic model based on the relativistic multiconfiguration Dirac–Fork (MCDF) method. We describe our model in detail in the next section. In section 3, we present the results of our calculation. The results obtained are discussed in the concluding section 4.

2. Method of calculations

In the present work the discrete $M1$ conversion in the thorium atom is treated within the framework of the relativistic MCDF method [7]. Computer codes for the MCDF calculations have been developed within our package of computer codes RAINE [8].

Generally, the probability of a discrete conversion transition from the ground atomic level with total angular momentum J_i of the initial state i to the final state f with the total momentum $J_f^{(n)}$ can be written as

$$W(J_i \rightarrow J_f^{(n)}) = \frac{2\pi}{\hbar}\frac{1}{2J_i+1}\sum_{M_i,M_f}\left|< \Psi(J_fM_f)|\hat{T}_q^L|\Psi(J_iM_i) >\right|^2. \tag{3}$$

Here $\Psi(JM)$ is the wave function of the atomic state with the momentum J and its projection M, $\hat{T}_q^L$ is the transition operator of the multipolarity L. It can be presented in terms of the sum of one-particle operators:

$$\hat{T}_q^L = \sum_{j=1}^{N}(\hat{t}_q^L)_j, \tag{4}$$

where N is the number of electrons in the atom.

Summation over the magnetic quantum numbers is usually performed by making use of the Wigner–Eckart theorem:

$$< \Psi(J_fM_f)|\hat{T}_q^L|\Psi(J_iM_i) > = \frac{1}{\sqrt{2J_f+1}}C^{J_fM_f}_{J_iM_iLq} < \Psi(J_f)||\hat{T}^L||\Psi(J_i) >, \tag{5}$$

with $C^{J_fM_f}_{J_iM_iLq}$ – the Clebsch–Gordan coefficients, and the reduced two-bar matrix element, which is no more depending on the magnetic quantum numbers.

By means of Equation (5), the probability (3) can be expressed in terms of the reduced matrix element:

$$W(J_i \rightarrow J_f^{(n)}) = \frac{2\pi}{\hbar}\frac{1}{2J_i+1}|< \Psi(J_f)||\hat{T}^L||\Psi(J_i) >|^2. \tag{6}$$

One can then obtain the value of the reduced matrix element in Equation (5) of the full atomic configurations in terms of the single-electron matrix elements, considering usual, one-bar matrix elements of the same transition operator for certain simple configurations with definite J and M. For this purpose, let us expand the full configuration atomic wave functions over the basis of Slater determinants $\Phi(M)$ with a certain projection M of the total angular momentum

$$\Psi(J_iM_i) = \sum_{\mu=1}^{N_i} c_{1\mu}\Phi_\mu(M_i)\,, \tag{7}$$

$$\Psi(J_f^{(n)}M_f) = \sum_{\nu=1}^{N_f} c_{n\nu}\Phi_\nu(M_f). \tag{8}$$

By substitution of the expansions (4), (7), (8) into Equation (3), the one-bar matrix element acquires the following form:

$$< \Psi(J_f^{(n)}M_f)|\hat{T}_q^L|\Psi(J_iM_i) >= \sum_{\mu,\nu} c_{1\mu}c_{n\nu} < \Phi_\nu(M_f)\left|\sum_{j=1}^{N}(\hat{t}_q^L)_j\right|\Phi_\mu(M_i) > . \tag{9}$$

In turn, the matrix element of the determinant wave functions for the one-particle operator takes the form [9]:

$$< \Phi_\nu(M_f)\left|\sum_{j=1}^{N}(\hat{t}_q^L)_j\right|\Phi_\mu(M_i) >=< \varphi(n_f\ell_fj_fm_f)|\hat{t}_q^L|\varphi(n_i\ell_ij_im_i) >< Y_\nu^f|Y_\mu^i > . \tag{10}$$

Here $\varphi(nljm)$ is the single-particle wave function of the 'active' electron involved in the transition, n is the principal quantum number, ℓ and j are the orbital and total angular momenta of the electron, m is the projection of the total momentum. The factor is the product of the overlapping integrals of the single-particle wave functions of all the electrons, with the exception of the 'active' one. This factor $< Y_\nu^f|Y_\mu^i >$ is either 1 or 0 if the single-electron wave functions are orthonormalized. It is different from 1 or 0 in the case if the mean field is different in the initial and final states: i.e., if one takes into account the effect of the hole produced in the final state as a result of internal conversion, or in calculations of the shake probability in internal conversion or beta decay, etc. Herein we use the orthonormalized set of the basis wave functions, and therefore, we shall not put down this factor explicitly in the formulae hereafter. Using then the Wigner–Eckart theorem again for the single-particle matrix element in Equation (10), and taking into account Equations (6)–(9), we arrive at the desired general expression for the probability of the discrete conversion transition:

$$W(J_i \to J_f^{(n)}) = \frac{2J_f^{(n)}+1}{2J_i+1}\left[C_{J_iM_iLq}^{J_f^{(n)}M_f}\right]^{-2}\left|\sum_{\mu,\nu,s,t} c_{1\mu}c_{n\nu}\frac{1}{\sqrt{2j_t+1}}C_{j_sm_sLq}^{j_tm_t}\mathcal{M}_{st}^{\tau L}\right|^2 . \tag{11}$$

In Equation (11), $\mathcal{M}_{st}^{\tau L}$ is the reduced matrix element of the 'active' electron with the quantum numbers $n_s l_s j_s$, to the state $n_t l_t j_t$. In our code, each of the valence electrons is tried as 'active', which results in the summation over the single-electron states s and t within each initial or final configuration, respectively.

We undertook calculations of the total conversion factor for the $M1$ transitions from the ground level of the initial state of the thorium atom Th(i) to all the allowed excited s or d states of the atom. Note that in our approach, they belong to the same set of the orthonormalized wave functions.

The required radiative width of a particular level has been evaluated in the same way, as a sum of the probabilities of the $E1$ transitions from this level to all the low-lying levels of the final atomic state Th(n). The following sets of the electron configurations were adopted for the initial and intermediate states:

$$\mathrm{Th}(i, l) = 7s^2 6d^2 + 7s6d^3 + 7s8s6d^2 + 7s9s6d^2 + 7s10s6d^2 \,, \tag{12}$$

whereas for the final p states, the following basis was used:

$$\mathrm{Th}(n) = 7s7p6d^2 + 7p6d^3 + 7s8p6d^2 + 7s9p6d^2 + 7s10p6d^2 \,. \tag{13}$$

All the possible spin-orbital combinations, within the framework of the $j-j$ coupling scheme, were taken into account. For example, $7s^2 6d^2 = 7s_{1/2}^2 6d_{3/2}^2 + 7s_{1/2}^2 6d_{3/2} 6d_{5/2} + 7s_{1/2}^2 6d_{5/2}^2$.

In the same way, the discrete conversion coefficient for the $M1$ transition from the ground level with $J_i = 2$ into the ℓ-th level $J_i^{(\ell)}$ can be obtained by means of Equation (9):

$$\alpha_d^{M1}(J_i \to J_i^{(\ell)}) = \frac{2J_i^{(\ell)} + 1}{2J_i + 1} \frac{1}{\left[C_{J_i M_i 1q}^{J_i^{(\ell)} M_i}\right]^2} \left| \sum_{\mu,\nu,s,t} c_{1\mu} c_{n\nu} \frac{1}{\sqrt{2j_t + 1}} C_{j_s m_s 1q}^{j_t m_t} \mathcal{M}_{st}^{M1} \right|^2 . \tag{14}$$

Here $\mathcal{M}_{st}^{M1}$ is the single-electron reduced conversion matrix element. It reads [3]

$$\mathcal{M}_{st}^{M1} = (-1)^{j-1/2} \left[\frac{1}{6} \alpha \pi \omega_n (2j_s + 1)(2j_t + 1)(2\ell_s + 1)(2\bar{\ell}_t + 1) \right]^{\frac{1}{2}} \times C_{\ell_s 0 \bar{\ell}_t 0} W\left(\ell_s j_s \bar{\ell}_t j_t ; \frac{1}{2} 1 \right) \times (\kappa_s + \kappa_t) \int_0^\infty (G_s F_t + F_s G_t) X_1(\omega_n r) dr \,. \tag{15}$$

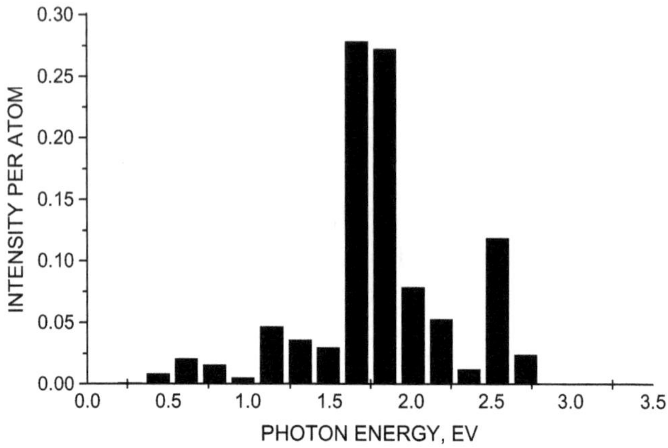

Figure 2. Calculated spectrum of the primary photons emitted after the resonance $7s \to 8s$ conversion. The converted electron makes a radiative transition to one of the lower-lying p levels.

In Equation (15), $X_1(\omega_n r)$ is the transition potential. We use the following expression for it, provided by the surface nuclear model for the transition currents [10], which takes a proper account of the finite nuclear size effects:

$$X_1(\omega_n r) = \begin{cases} j_1(\omega_n r)\dfrac{h_1(\omega_n R_0)}{j_1(\omega_n R_0)} & \text{if} \quad r \leq R_0, \\ h_1(\omega_n r) & \text{if} \quad r > R_0\,. \end{cases} \tag{16}$$

In Equations (15) and (16), $\bar{\ell} = 2j - \ell$, $\kappa = (\ell - j)(2j + 1)$ is the relativistic quantum number, $W(\ell_s j_s \bar{\ell}_t j_t; \frac{1}{2}\ 1)$ is the Racah coefficient, G and F are the large and small components of the electron radial wave function multiplied by r, α is the fine structure constant, R_0 is the nuclear radius, $\omega_n = 3.5$ eV is the nuclear transition energy, $j_\Lambda(\omega_n r)$ and $h_\Lambda(\omega_n r)$ are the spherical Bessel and Hankel functions, respectively. The formulae are given using the relativistic units: $\hbar = c = 1$.

The total conversion factor can then be obtained by summation over all the levels ℓ, allowed by the selection rules for the $M1$ transition, as follows:

$$R_{tot} = \sum_{\ell} R(J_i \to J_i^{(\ell)})\,. \tag{17}$$

The width $\Gamma^{(\ell)}$ in Equation (2) is defined by the sum of the probabilities of the electric dipole $E1$ radiative transitions from the ℓ-th level of the intermediate state Th(l) (Figure 1) to all the lower-lying levels of the final state Th(n):

$$\Gamma^{(\ell)} = \sum_{n} W(J_i^{(\ell)} \to J^{(n)})\,. \tag{18}$$

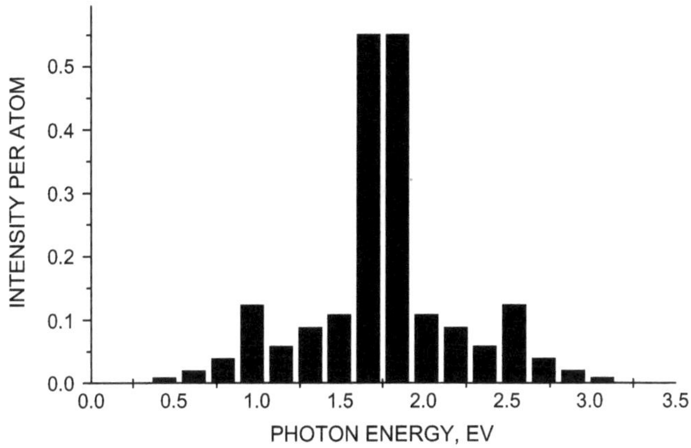

Figure 3. Calculated total (symmetric) spectrum of the delayed photons emitted after the resonance $7s \rightarrow 8s$ conversion. The converted electron accomplishes the cycle through the chain of radiative transitions $8s \rightarrow 7p \rightarrow 7s$ (Figure 1b), returning to its initial state.

The probability $W(J_i^{(\ell)} \rightarrow J^{(n)})$ is also found by making use of Equation (11). The expression for the corresponding matrix element $\mathcal{M}_{st}^{(E1)}$ in the length gauge, used in the present calculation, reads as follows:

$$\mathcal{M}_{st}^{(E1)} = (-1)^{\ell_s+\ell_t}[\alpha k(2j_t+1)(2j_s+1)(2\ell_s+1)(2\ell_t+1)]^{1/2} C^{10}_{\ell_s 0 \ell_t 0} W\left(\ell_s j_s \ell_t j_t; \frac{1}{2}\, 1\right)$$
$$\times \int_0^\infty \{2(G_s G_t + F_s F_t) j_1(kr) + [(\kappa_t - \kappa_s - 2) G_s F_t + (\kappa_t - \kappa_s + 2) G_t F_s] j_2(kr)\} dr\,. \tag{19}$$

with $k = \omega_n - \epsilon_n$ being the transition energy.

3. Results

The calculated spectrum of the primary photons with the energies ω_1, resulting from the $8s - 7p$ transition, is presented in Figure 2. It was calculated for the nuclear energy $\omega_n = 3.5$ eV.

In Figure 3, we present the full spectrum calculated under assumption that the rest of the nuclear energy is brought off by the other photon with the energy $\omega_2 = \omega_n - \omega_1$ (Figure 3). For the total R factor, the value of $R = 4788$ was obtained, which is in agreement with that reported earlier [2].

4. Discussion

As one can see from the results of the previous section, the resonance conversion appears to be the predominant decay channel. This has important consequences for experimental research.

Firstly, resonance conversion very radically reduces the lifetimes, from tens hours to a few minutes or even less. Moreover, in the case of accidental coincidence of the nuclear and atomic frequencies, the lifetime may be reduced up to 10^{-11} s [6].

Secondly, the results clearly demonstrate that it makes not much sense to look for direct photons from the radiative nuclear decay. Experimental efforts have to be focussed on search for delayed photons coming from the resonance bridges, with the energies approximately half the nuclear energy (Figure 3).

These conclusions radically affect prospects for observing any other decay mode. Thus, in the case of α decay, the resonance conversion not only reduces the lifetime, as shown previously, but also reduces the probability itself of α decay by approximately R times. Indeed, for a time of observation t_{obs} more than the total lifetime, $t_{obs} > \frac{\hbar}{\Gamma^{(t)}}$, due to the effect of resonance conversion, the number of alphas observed also diminishes by approximately R times.

Summing it up, we conclude that resonance appears to be the absolutely predominant decay channel. Experimental efforts at the contemporary stage of investigation, therefore, should be aimed at search for delayed soft photons with the energies of about 1.5–2eV, resulting from the decay of the isomer via resonance bridges.

Acknowledgements

This work was supported by DTRA (USA) under contract *No. DTRA*; 01–02–M–0534, and by Russian Foundation for Basic Research under grant *No*. 05-02-17340.

References

1. Karpeshin F. F., Band I. M., Trzhaskovskaya M. B. and Listengarten M. A., *Phys. Lett.* **B372** (1996), 1.
2. Karpeshin F. F., Band I. M., Trzhaskovskaya M. B. and Pastor A., *Phys. Rev. Lett.* **83** (1999), 1072; Izv. Rossiisk. Akad. Nauk, Ser. Fiz. **63** (1999), 38. (*English transl.* Bull. Russian Acad. Sci., Phys. Ser. (USA) **63** (1999), No. 1).
3. Karpeshin F. F., Harston M. R., Attallah F., Chemin J. F., Scheurer J. N., Band I. M. and Trzhaskovskaya M. B., *Phys. Rev.* **C53** (1996), 1640.
4. Zaretsky D. F. and Karpeshin F. F., *Sov. J. Nucl. Phys.* **29** (1979), 151.
5. Karpeshin F. F., *Hyperfine Interact.*, **143** (2002), 79.
6. Gangrsky Yu. P., Zemlyanoi S. G., Zhemenik V. I., Karpeshin F. F., Myshinsky G. V. and Trzhaskovskaya M. B., Izv. Rossiisk. *Akad. Nauk, Ser. Fiz.* **69** (2005), 1663.
7. Grant I. P., McKenzie B. J., Norrington P. H., Mayers D. F. and Pyper N. C., *Comput. Phys. Commun.* **21** (1980), 207.
8. Band I. M. and Trzhaskovskaya M. B., *ADNDT* **55** (1993), 43.
9. Bethe H. A., *Intermediate Quantum Mechanics*, W.A. Benjamin, Inc., N.-Y.-Amsterdam, 1964, p. 60.
10. Sliv L. A., *Zh. Exp. Teor. Fiz.* **21** (1951), 770.

Hyperfine Interactions (2005) 162:133–138
DOI 10.1007/s10751-005-9214-0

Study of Exotic Nuclei

M. HEMALATHA[1,*], S. KAILAS[1] and Y. K. GAMBHIR[2]
[1]*Nuclear Physics Division, Van de Graaff Building, Bhabha Atomic Research Centre, Mumbai 400 085, India; e-mail: mhema@magnum.barc.ernet.in*
[2]*Department of Physics, IIT-Powai, Mumbai 400 076, India*

Abstract. A high-sensitive fluorescence cell has been developed with an aim to perform laser spectroscopy of exotic nuclei. This fluorescence cell has been tested off-line for stable isotope ^{133}Cs. Also, an investigation of the nuclear root mean square (r.m.s.) charge and neutron radii, and of the binding energies of the cesium long isotopic chain has been carried out in the relativistic mean field (RMF) and relativistic Hartree–Bogoliubov (RHB) formalisms. The RMF/RHB calculations are compared with the experimental data and are found to be in good agreement.

Key Words: cesium isotopic chain, exotic nuclei, laser spectroscopy, nuclear r.m.s. radii, relativistic mean field theory.

1. Introduction

The physics of exotic nuclei is an emerging frontier in nuclear physics and allied sciences [1]. It is expected to address the basic questions regarding the nature of nuclear interaction, the origin of elements in the universe and the role of isospin in determining the structure and properties of nuclei. Exotic nuclei are nuclei lying away from the line of β-stability, having an extreme neutron number to proton number (N/Z) composition and a relatively short half-life.

With the development of the on-line mass separators coupled to powerful accelerators, a wide range of short-lived nuclei far off stability became accessible for investigation. But most of these nuclei are available in trace quantities only and for a limited amount of time. Hence the usual scattering and nuclear reaction experiment using these nuclei as target or beam is difficult to perform. The low intensities of radioactive ion beams available at present coupled with their short lifetimes require special measurement techniques to be employed for exotic nuclei.

The high sensitivity and resolution of laser spectroscopy gives it a unique role in the investigation of ground and isomeric states of nuclei far from the line of stability. Precise and systematic studies of the isotope shifts (IS) of optical

* Author for correspondence.

transitions provide the information on changes in the mean-square nuclear charge radii and hyperfine structures (hfs) give information about the multipole moments [2–4]. High sensitivity is important since these nuclei are produced in trace quantities while high resolution makes it possible to detect small differential changes in the mean-square charge radii between the isotopes. The description of the nuclear radii systematics presents a substantial challenge to even the most sophisticated nuclear models. In order to understand the systematics of ground state properties of Cs isotopic chain, we carry out the investigation using Relativistic Mean Field (RMF) and Relativistic Hartree–Bogoliubov (RHB) formalism.

In this talk, the production of neutron-deficient nuclei ^{127}Cs and ^{129}Cs by heavy-ion fusion reaction required for laser spectroscopy and the development of a high-sensitive fluorescence cell for the measurement of atomic isotope shift and hyperfine structure using laser spectroscopy will be outlined. This will be followed by the systematic investigation of Cs isotopic chain by RMF and RHB formalism.

2. Production of neutron-deficient nuclei ^{127}Cs and ^{129}Cs by heavy-ion fusion reaction

As part of our program of measurement of ground state properties of radioactive nuclei by laser spectroscopy, we have measured the production yields of neutron-deficient nuclei ^{127}Cs and ^{129}Cs in the reaction ^{11}B with ^{nat}Sn target using heavy-ion fusion reaction [5]. The production rates of ^{127}Cs and ^{129}Cs were measured using a stacked-foil activation technique. The half-lives of these isotopes being long, radioactivity of the activation products was determined via off-line γ-counting method. The stack consisting of ^{nat}Sn target and Al was exposed to ^{11}B beam of energy 52 MeV, from 14UD BARC-TIFR Pelletron accelerator facility at Mumbai, for about 10 h at a current of $\sim$ 5 nA.

It is estimated that there must be $\sim 10^4$ atoms in the interaction region for the laser spectroscopic experiments to be done. A ^{nat}Sn target of thickness of $\sim$ 20 mg/cm^2 irradiated for saturation activity with 52 MeV ^{11}B beam (current scaled to 1μA) results in a production of ^{127}Cs and ^{129}Cs of 10^7 and 10^8 nuclei per second, respectively. The Cs nuclei produced in Sn target/Al catcher foil will be separated using radiochemical techniques and converted as nitrate for laser spectroscopic work.

3. Development of a high-sensitive fluorescence cell for the measurement of atomic IS and Hfs using laser spectroscopy

We report our work on the development of a high-sensitive laser induced fluorescence (LIF) set-up for measurement of IS and hfs of exotic isotopes. LIF spectroscopy has been specifically chosen for its sensitivity and versatility. To

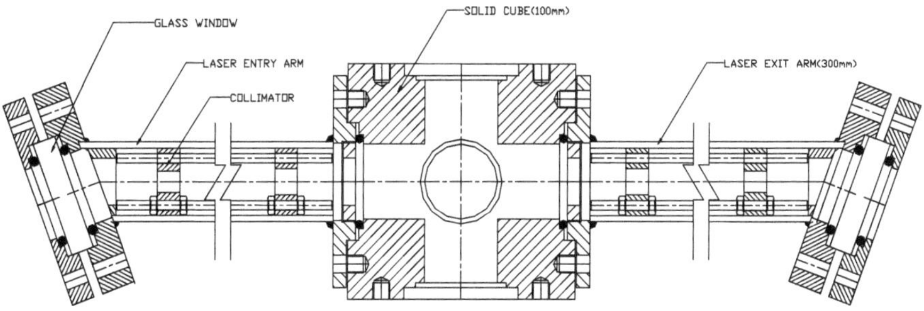

Figure 1. Schematics of the fluorescence cell.

carry out the laser spectroscopy work, a versatile fluorescence cell, which can be used for on-line as well as off-line experiments, has been designed and fabricated. While off-line experiments can be done with nuclei having long half-lives (> 1 day), the on-line experiments are aimed to exploit the new production possibilities offered by accelerated radioactive ion beam facilities.

A schematic representation of the fluorescence cell is shown in Figure 1. It consists of a vacuum chamber having six ports on six faces of a cube. Four of the ports are, respectively, used for entry of the ion beam into the chamber, for fluorescence collection and detection, and for entry of buffer gas with LN_2 cooling. The two remaining horizontal ports allow for laser entry and exit. They have extended arms with collimators to reduce the contribution of scattered laser light. These arms are provided with optical windows at a small angle to prevent entry of the reflected light in the region of fluorescence detection. For off-line experiments, a small heater has been designed which can be used in place of the ion beam entry. The heater assembly is essentially a small oven to form an atomic beam. Standardization of the set-up is being done in an off-line mode. The laser used in this study is an external cavity single mode, with a line width of 1 MHz, tunable diode laser with centre frequency 852 nm which can address 6s $^2S_{1/2}$ $F = 4 \rightarrow$ 6p $^2P_{3/2}$ $F = 5$ hyperfine transition in the isotopic chain of Cs nuclei. Since the production rates of many of the isotopes are in the range of 10^4 to 10^6 atoms/s, the detection technique must be sensitive enough to reach this limit. Our initial experiments on fluorescence detection using a cesium vapour cell are shown in Figure 2. In these experiments the diode laser is locked to the 6s $^2S_{1/2}$ $F = 4 \rightarrow$ 6p $^2P_{3/2}$ $F = 5$ hyperfine transition of cesium and the fluorescence emanating from the cesium vapour source is detected using collection optics and a red-enhanced photomultiplier tube.

It may be noted that the detection limit is governed by the background contributed mainly by the scattering of the laser at the windows and walls of the cell. In the resonance cell described above, we have incorporated various features, which ensure very high sensitive detection.

Since the radioactive Cs obtained is in the form of cesium nitrate, it is dispersed in titanium powder and heated to about 400–500°C, to form Cs atoms. LIF in collimated atomic beam technique was used and fluorescence is monitored

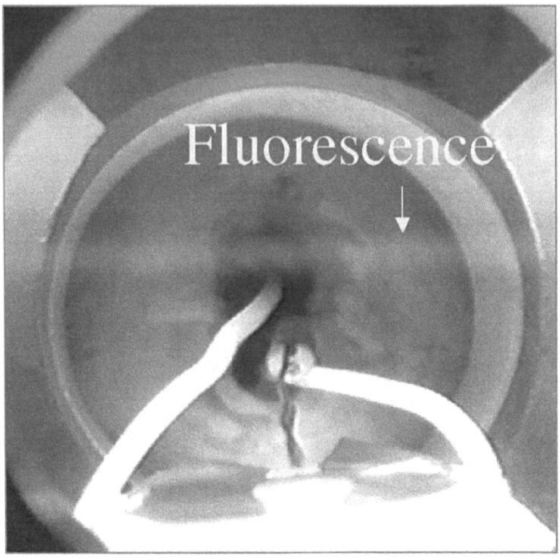

Figure 2. Picture showing fluorescence in the fluorescence cell.

using an infrared CCD camera. The sample is implanted in the oven and subsequently evaporated into the vacuum. An atomic beam of Cs atoms was formed by collimation and irradiated by the diode laser. Laser induced fluorescence in collimated atomic beam was carried out and the atomic fluorescence was recorded using a CCD camera as shown in Figure 2.

A fluorescence cell has been designed for on-line experiments at accelerators and has been tested off-line using stable isotope ^{133}Cs. Efforts are underway to determine the sensitivity of the set-up and improve it so that nuclei (stable and radioactive) in small concentrations can be studied.

4. Systematics of charge radii of Cs isotopes using RMF

The isotopic shift measurements by laser spectroscopy of Cs isotopic chain [6] reveal a rich structure (Figure 3(b)). For example, there is a "kink" at $N = 82$ and the radii deviate from the conventional $r_0A^{1/3}$ rule considerably as shown in Figure 3. The decrease in charge radii with decreasing neutron number is extremely slow in the neutron-deficient region, followed by a jump at 65–64 and then a sudden fall from 64–63. This feature is supposed to be a result of valence level structure near the Fermi surface. It is, therefore interesting to carry out a systematic investigation of ground state properties of Cs isotopes.

The RMF/RHB equations have been solved employing the basis expansion (spherical or deformed oscillator basis) method. The results of RMF calculations with frozen gap approximation are denoted by SPH. RHB using finite range Gogny D1S interaction are denoted by RHB(ob). RMF equations with constant gap approximation in the deformed oscillator basis with axial symmetry are denoted by DEF. More details can be found in [7, 8].

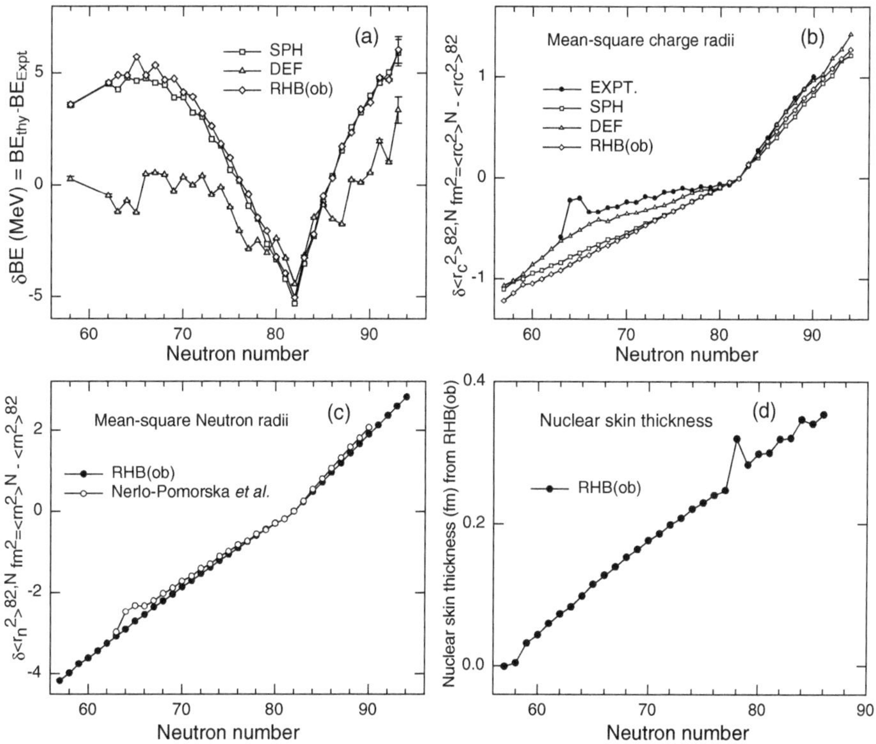

Figure 3. (a) Difference in the calculated and experimental binding energies. (b) The change in the root mean square charge radii relative to $N = 82$ (^{137}Cs) as a function of neutron number. (c) The change in the root mean square neutron radii relative to $N = 82$ plotted as a function of neutron number. (d) Nuclear skin thickness as a function of neutron number.

The differences in the binding energies calculated by RMF/RHB equations and the corresponding experimental values [9] are shown in Figure 3(a). It is clear from figure, that the inclusion of deformation brings the calculations closer to the experiment.

The change in the r.m.s. charge radii relative to $N = 82$ (^{137}Cs) for the chain of Cs isotopes obtained in RMF/RHB calculations as well as from experimental data [6] is shown in Figure 3(b). There is a little odd-even staggering for neutron-deficient nuclei whereas there is no staggering for neutron-rich nuclei. Experimentally, there is a large "kink" between $N = 63$–66. The DEF calculations, where deformations are taken into account, agree with the experiment except between $N = 63$–66.

The change in the r.m.s. neutron radii relative to $N = 82$ plotted as a function of neutron number is shown in Figure 3(c). The calculated r.m.s. radii agree remarkably well with Nerlo-Pomorska and Mazurek [30] which are calculated from a proposed algebraic expression of RMF neutron radii based on experimental neutron radii extracted from *pA* scattering by Batty *et al.* [11].

One of the very interesting quantities in nuclear structure study is the nuclear skin: Difference between r.m.s. neutron and proton radii. Calculated nuclear skin thickness as a function of neutron number is displayed in Figure 3(d). Clearly, there is a monotonous increase for large nuclear skin at N = 82 which is a reflection of closed shell structure. One expects a strong correlation between the nuclear skin thickness and the corresponding difference between the single neutron and proton separation energies. Indeed there is a strong negative correlation. The degree of correlation is quantified by carrying a linear regression analysis.

The RMF description of the Cs isotopes agrees remarkably well with the experimental data available, except for N = 64–66 region.

Acknowledgements

One of the authors (MH) would like to express her sincere gratitude to Prof. H.-J. Kluge for providing financial support for attending this workshop. MH also acknowledges the fellowship from Department of Atomic Energy of the Government of India.

References

1. Casten R. F. and Sherrill B. M., *Prog. Part. Nucl. Phys.* **45** (2000), S171.
2. Otten E. W., *Treatise on Heavy-Ion Science*, Vol. 8, Ed. D. A. Bromley Plenum, New York, 1989.
3. Kluge H.-J., *Spectroc. Acta* **B 58** (2003) 1031.
4. Billowes J. and Campbell P., *J. Phys., G, Nucl. Part. Phys.* **21** (1995), 707.
5. Hemalatha M. *et al., Nucl. Instrum. Methods* **A531** (2004), 645.
6. Thibault C. *et al., Nucl. Phys.* **A367** (1981), 1.
7. Gambhir Y. K. *et al., Ann. Phys.* **198** (1990), 132.
8. Kucharek H. and Ring P., *Z. Phys.* **A339** (1991), 23.
9. Audi G. *et al., Nucl. Phys.* **A729** (2003), 337.
10. Nerlo-Pomorska B. and Mazurek K., *arXiv nucl-th/0211049, and Phys. Rev.* **C66** (2002), 064305.
11. Batty C. J. *et al., Adv. Nucl. Phys.* **19** (1989), 1.

Hyperfine Interactions (2005) 162:139–146
DOI 10.1007/s10751-005-9216-y

Calculations of the Hyperfine Anomaly in the Lanthanides

J. R. PERSSON
Kristianstad University, SE-29188 Kristianstad, Sweden; e-mail: Jonas.persson@mna.hkr.se

Abstract. Calculations of the Bohr–Weisskopf effect and hence the hyperfine anomaly has been performed in Eu using the particle-rotor formalism. Comparisons with experimental data and a discussion on the semi-empirical Moskowitz–Lombardi formula is performed.

1. Introduction

The influence of the finite size of the nucleus on the hyperfine structure (hfs) was first considered by Bohr and Weisskopf [1]. They calculated the hyperfine interaction (hfi) of $s_{1/2}$ and $p_{1/2}$ electrons in the field of an extended nucleus, and showed that the magnetic dipole hyperfine interaction constant (A) for an extended nucleus is generally smaller than that expected for a point nucleus. Isotopic variations of magnetic moments become larger than those in the point dipole interaction when there are different contributions to the hfs from the orbital and spin parts of the magnetisation in the case of extended nuclei. The fractional difference between the point nucleus hfi constant (A_{point}) and the constant obtained for the extended nuclear magnetisation is commonly referred to as the Bohr–Weisskopf (BW) effect [2]. The hfi constant A can therefore be written as

$$A = A_{point}(1 + \epsilon_{BW}) \tag{1}$$

where ϵ_{BW} is the BW-effect, and A_{point} the A constant for a point nucleus. The BW-effect is dependent on both a nuclear part as well as an atomic part, i.e. the electron density within the nucleus. The nuclear part, i.e., the distribution of nuclear magnetisation, can be calculated using different nuclear models [2, 3]. Because electronic wave functions cannot be calculated with high accuracy in complex atoms, as they can be in hydrogen-like ions and muonic atoms, it is not always possible to determine ϵ_{BW} directly. However, it is possible to determine the difference of the BW-effect in two isotopes, the so-called (differential) hyperfine anomaly (hfa). Where one compares the ratio of the measured hfs constants for two isotopes with the independently measured ratio of the nuclear

Table I. BW-effect from muonic atoms

Isotope	ϵ_{BW}
^{139}La	−0.156(46)%
^{141}Pr	−0.383(23)%
^{151}Eu	−0.63(13)%

magnetic dipole moments to extract the hfa, $^1\Delta^2$, for the isotopes 1 and 2, and a given atomic state:

$$1 + {}^1\Delta^2 = \frac{A^{(1)}\mu_I^{(2)}/I^{(2)}}{A^{(2)}\mu_I^{(1)}/I^{(1)}} \approx 1 + \epsilon_{BW}^{(1)} - \epsilon_{BW}^{(2)} \tag{2}$$

For electrons with a total angular momentum $j > 1/2$ the anomalies may be disregarded as the corresponding wave functions vanish at the nucleus. The hfa can show a dependence of the atomic state, a state dependent hfa, where the values for different states can vary significantly. The reason is that the hyperfine interaction consists of three parts [4, 5], orbital, spin-orbit and contact (spin) interaction, where only the contact interaction contributes to the hfa. It is suitable to rewrite the dipole hyperfine interaction constant as

$$A = A_{nc} + A_c \tag{3}$$

where A_c is the contribution due to the contact interaction of s (and $p_{1/2}$) electrons and A_{nc} is the contribution due to non-contact interactions. The experimental hfa, which is defined with the total magnetic dipole hyperfine constant A, should then be rewritten to obtain the relative contact contribution to the hfa:

$${}^1\Delta^2_{exp} = {}^1\Delta^2_c \frac{A_c}{A} \tag{4}$$

where ${}^1\Delta^2_c$ is the hfa due to the contact interaction, that is, for an s- or p-electron. So far we have considered direct interactions between the electron and the nucleus, but we should also include electron–electron interactions. One interaction, which can influence the hyperfine interaction, is the polarisation of the electron core [4], which may give a substantial contribution to the experimental hfa [2]. Core polarisation can be seen as an excitation of a d-electron, which do not give any contribution to the hfa, to an s-electron, which gives a large hfa. Since ${}^1\Delta^2_s$, is independent of n, it is possible to use it to obtain values of the core-polarisation [2, 6].

From the discussion, one is led to the conclusion that one need independent measurements of the nuclear magnetic moment and the A-constant in order to obtain the hfa, this is not true. It has been shown by Persson [7] that it is possible to extract the anomaly solely from the A-constants, provided the ratio

Table II. Hyperfine anomaly in the lanthanides

Isotopes	${}^{1}\Delta_{s}^{2}$
^{143,145}Nd	0.2034(63)%
^{151,153}Eu	−0.64(3)%
^{155,157}Gd	−0.106(24)%
^{171,173}Yb	−0.386(5)%

$(\frac{A_s}{A})$ differs substantially for the A-constants. Comparing the A-constants ratio, for two isotopes, in two atomic states, gives:

$$\frac{A_B^{(1)}/A_B^{(2)}}{A_C^{(1)}/A_C^{(2)}} \approx 1 + {}^{1}\Delta_s^2\left(\frac{A_s^B}{A^B} - \frac{A_s^C}{A^C}\right) \tag{5}$$

Where B and C denotes different atomic levels and 1 and 2 denotes different isotopes. The ratio between the two-constant ratios for the isotopes will only depend on the difference of the contact contributions of the two atomic states and the hfa for the s electron. It should be noted that the ratio $(\frac{A_s}{A})$ is isotope independent. Once determined for one isotopic pair, the ratio can be used for all pairs, which is useful in the study of hfa in radioactive isotopes. The ratio can be determined in two different ways; either by making an analysis of the hyperfine interaction and deducing from the analysis, or by using a known hfa as a calibration. It should be pointed out that the atomic states used must differ in the ratio$(\frac{A_s}{A})$, as a small difference will lead to an increased sensitivity to errors, as can be deduced from Equation (5).

In order to determine the hfa, one needs to know either the nuclear magnetic dipole moment and one A-constant or two A-constants [2, 7]. Since the hfa is normally very small (1% or less) it is necessary to have high accuracy, better than 10^{-4} [2]. In the case of stable isotopes there is no major problem to measure the nuclear magnetic moment, with NMR or ABMR, while unstable isotopes are more difficult. In most cases there does not exist any high precision measurements of the nuclear magnetic moment. However, there might exist measurements of two A-constants, if the unstable isotopes nuclear charge radius has been measured by means of laser spectroscopy [8]. In order to obtain the hfa one needs to measure the A-constants with an accuracy better than 10^{-4}. This can be done by laser spectroscopy when the A-constant is larger than 1,000 MHz.

2. Hyperfine anomaly in the lanthanides

The lanthanides have been an area of interest from both an atomic and a nuclear physics view. Even if there has been done systematic studies of the atomic structure as well as the isotope shift, the BW-effect and hfa have not been well studied. This despite the fact that one of the largest state dependent hfa has been

Table III. Hyperfine anomaly in Eu, experimental and calculations

	$^{151}\Delta_s^A(\%)$	$^{151}\Delta_{MO}^A(\%)$	$^{151}\Delta_I^A(\%)$ [14]	$^{151}\Delta_{II}^A(\%)$ [14]	$^{151}\Delta_{ML}^A(\%)$
	Exp.	Calc.	Calc.	Calc.	Calc.
^{145}Eu	−0.08(15)	0.000	0.021	0.031	0.056
^{147}Eu	−0.12(17)	0.002	0.010	0.008	0.029
^{149}Eu	−0.19(16)	0.002	0.007	0.006	0.011
^{151}Eu	0	0.0	0	0	0.000
^{153}Eu	−0.64(4)	−0.768	−0.127	0.003	−0.546
^{155}Eu	−0.91(37)*	−0.768	−0.127	−0.001	−0.555

* state dependent anomaly.

observed in Eu [9, 10] One reason for this might be that the nuclear magnetic moments are not known with high accuracy, another being the rather complicated atomic structure. The BW-effect has been deduced from muonic atoms in three of the lanthanides, making it possible to extract the BW-effect for a series of isotopes and make direct comparisons with calculations. The BW-effect for muonic atoms is given in Table I [2].

The hfa has been extracted in four elements and isotope pairs. As we are only considering the s-electron anomaly here, we exclude the state dependent hfa that has been obtained in some of the lanthanides. The values of the s-electron anomaly are given in Table II [2, 6]. We observe that Eu is a special case, where the BW-effect as well as the hfa is known. In addition, there have been done measurements of the hfs in a long chain of isotopes [11–13], making it possible to use the method of Persson [7] to extract the hfa.

It should be noted that the A-constants for two atomic states in Eu^+ has been measured in an ion-trap, both for the stable isotopes as for the unstable ^{148}Eu and ^{150}Eu [19–21]. This would, in principle, allow for an determination of the hfa using these results, but this is not the case. The studied states, the ionic ground state $^9S_4$ and the metastable $^7S_3$, are dominated by the contact interaction in such a way that the ratio $\left(\frac{A_s}{A}\right)$ differs very little between the states leading to an large uncertianty, as discussed earlier, making them unuseful for this purpose. Neither is it meaningful to try and use them with the other results [11–13] as the hfa is deduced using the known hfs for the stable Eu isotopes as a calibration, nor is an analysis of the hfs in the states involved possible.

The hfa has been deduced for all odd isotopes from $A = 145$ to $A = 155$ using the known hfa of ^{151,153}Eu as a calibration, [6] and the result is given in Table III. Note that the experimental value for ^{155}Eu is state-dependent and not the s-anomaly.

The shape transition between ^{151}Eu and ^{153}Eu drastically changes the anomaly. The hfa for the lighter isotopes is fairly constant, indicating that the magnetisation does not change much from the spherical ^{145}Eu nucleus to ^{151}Eu. With

Table IV. Magnetic moments and hyperfine anomalies for odd Eu isotopes

A	$\mu_I(exp)$	μ_I(MO)	ϵ(%)(MO)	ϵ(%) I [14]	ϵ(%)II [14]
145	3.993	3.773	−1.001	−1.067	−1.067
147	3.724	3.720	−1.003	−1.056	−1.044
149	3.565	3.552	−1.003	−1.053	−1.042
151	3.4717	3.506	−1.004	−1.046	−1.036
153	1.5330	1.532	−0.236	−0.919	−1.039
155	1.52	1.529	−0.236	−0.919	−1.035

MO denotes values obtained from particle-rotor calculations.

these experimental results we can make comparisons with theoretical calculations. The experimental results are in Table III compared with the results of Asaga *et al.* [14] (I, II), who used a microscopic model with core-excitations to calculate the BW-effect and with results from particle-rotor calculations based on the modified oscillator (MO) potential. The hfa has also been calculated using empirical Moskowitz–Lombardi (ML) formula [16], where the constant α was taken to be 0.015 nm. As can be seen all methods reproduce the trend. However, it seems like the particle-rotor calculations are better. Both theoretical methods and the ML formula indicate that the lighter isotopes have the same BW-effect as ^{151}Eu.

3. Calculations of the hyperfine anomaly

The BW-effect, and thereby the hfa, is investigated by making particle-rotor calculations based on the modified oscillator (Nilsson) potential, using standard parameters [15] as far as possible. As the nuclear magnetic moment and the BW-effect calculations are mainly analogous, it is sensible to adjust the parameters in the calculation so that both the energy levels and nuclear magnetic moments fits well with experimental values. The calculated BW-effect in the odd isotopes are given in Table IV. As expected the BW-effect, stays fairly constant from $A = 145$ to $A = 151$ with an abrupt change at the shape transition between $A = 151$ and 153.

There is a significant difference in the values obtained from the two methods of calculation. As we have an experimental value of the BW-effect in ^{151}Eu from muonic X-ray measurements $\epsilon_{BW} = -0.63(13)\%$ it is possible to further discuss the methods. It is clear that the particle-rotor calculations show a better agreement with experiment. The agreement with experiment in Eu might be a coincidence, as preliminary calculations in Gd and Sm shows a more complex situation [6].

Table V. Hyperfine anomaly in the lanthanides

	$\mu_{I,1}$	$\mu_{I,2}$	${}^{1}\Delta_{s}^{2}(\%)$	$\alpha(10^{-2})$
${}^{143,145}_{60}$Nd	−1.065	−0.656	0.2034	0.35
${}^{151,153}_{63}$Eu	3.4717	1.533	−0.64	1.76
${}^{157,155}_{64}$Gd	−0.3387	−0.2572	−0.106	0.11

4. The empirical Moskowitz–Lombardi formula

The empirical ML formula was established in 1973 as a rule for the s-electron BW-effect in mercury isotopes [16].

$$\epsilon_{BW} = \frac{\alpha}{\mu_I}, \alpha = \pm 1.13 \cdot 10^{-2} \mu_N \text{ for } I = l \pm \frac{1}{2} \tag{6}$$

where l is the orbital momentum for the odd neutron. It turned out that the empirical rule provided a better agreement with experimental hfa than the theoretical calculations performed by Fujita and Arima [3] using microscopic theory. The rule can be qualitatively explained by the microscopic theory used by Fujita and Arima [3], where the parameter α is more state independent than given by the theory. Further investigations gave an analogous expression for the odd-proton nuclei ^{191,193}Ir, ^{197,199}Au and ^{203,205}Tl, but also for the doubly odd ^{196,198}Au nuclei. The results indicate that the spin operators $g_s^{(i)}\Sigma_i^{(1)}$ are state independent for these nuclei. It is worth noting that all nuclei discussed lie close to the doubly closed shell nucleus ^{208}Pb, where one would expect the single particle model to provide a good description of the nucleus. It is not apparent that the rule is applicable to lighter nuclei.

With the data presented here it is possible to make a comparison with lanthanide nuclei. As has been shown in Eu, the ML formula seems to account for the hfa, even though the obtained value of the BW-effect differs more from the experimental value of ^{151}Eu. It should be noted that the sign of α is different from the value obtained for nuclei close to ^{208}Pb, indicating that the ML rule is not universal. In order to further test the ML formula, the values of α was deduced from the experimental values of the hfa and nuclear magnetic moments in Nd, Eu and Gd, using:

$${}^{1}\Delta^{2} = \alpha\left(\frac{1}{\mu_{I,1}} - \frac{1}{\mu_{I,2}}\right) \tag{7}$$

If the ML rule show some sort of general validity the values of α should stay fairly constant and show a different sign between Eu (odd-proton) and Nd and

Gd (odd-neutron). The obtained values are shown in Table V. As can be seen there are no indications that the ML rule is applicable for these nuclei. The conclusion would be that one should be very careful in applying the ML rule for lighter nuclei.

5. Discussion & conclusion

As has been shown in the case of Eu, it is possible to obtain a lot of information on the hfa without knowing the nuclear magnetic moment of the isotopes under study. The lanthanides are characterised by a vast number of low-lying energy states [17]. A general feature is the small values of the A-constants in the low lying levels of the $4f^n6s^2$ and $4f^{n-1}5d6s^2$ configurations, where neither free s-electron nor core-polarisation contribute to the hyperfine interaction [18]. In other words we have a number of states where the hfa will be practically zero. These states are therefore suitable in determining the ratio of the nuclear magnetic dipole moment. However, the A-constants in these states are in most cases small ($\leq$ 100 MHz) so the accuracy obtained by laserspectroscopy is normally not sufficient for determining the hfa, why the use of rf-spectroscopy is preferred. If we cannot make use of the zero anomaly states, we need to find at least 2 states where the ratio $\left(\frac{A_s}{A}\right)$ differs substantially. Since $\left(\frac{A_s}{A}\right)$ is proportional to $(g_J - 1)$, we can easily find suitable candidates. The best choice is the lowest J-value and one or two of the highest J-values within a multiplet. One caveat is that the states, should be close to pure LS-coupling. This has been found to be the case in many of the lowest lying multiplets in both atoms and ions of the lanthanides. As the number of states with excitation energy $\leq$10,000 cm^{-1} is high in most of the lanthanides, there are plenty of suitable transitions that can be studied with simple atomic beam sources. However, in order to populate other states than the ground state, one can easily loose signal. An alternative would be to combine rf-measurements of the atomic ground state, which normally has a small A-constant, with measurements of the ground state A-constant in the ion.

The ground states of singly charged lanthanide ions normally belong to a configuration with an un-paired s-electron. One might also consider doing ion-trap experiments using both singly and doubly charged ions, again in order to have one state with an unpaired s-electron and one without. The measurements of the hfs of Eu^+ (both stable and unstable) [19–21] shows that this is possible. As the prospect for systematic measurements of the hfa in the lanthanides involves more than one experimental technique, it seems likely that an international collaboration would be preferable. The conclusion would be to form a network of laboratories, in order to promote the study of hfa. As a start a Web-based database with experimental values has been created [http://www.mna.hkr.se/~pej/hfa.html], to which researchers are invited to submit their results. It seems feasible that, within a not too distant future, the hfa is well known and, hopefully, well understood in the lanthanides.

It has also been shown that the ML rule is not universal, why one has to be careful in applying it to nuclei far from the doubly closed shell nucleus ^{208}Pb.

References

1. Bohr A. and Weisskopf V. F., *Phys. Rev.* **77** (1950), 94.
2. Büttgenbach S., *Hyperfine Int.* **20** (1984), 1.
3. Fujita T. and Arima A., *Nucl. Phys.* **A254** (1975), 513.
4. Lindgren I. and Morrison J., *Atomic Many-Body Theory*, Springer, Berlin Heidelberg New York, 1983.
5. Sandars P. G. H. and Beck J., *Proc. R. Soc. London* **A289** (1965), 97.
6. Persson J. R., in manuscript.
7. Persson J. R., *Euro. Phys. J.* **A2** (1998), 3.
8. Otten E. W., In: D. Allan Bromely (ed.), *Treatise on Heavy-Ion Science, vol. 8,* 1989, p. 515.
9. Eliel E. R. *et al., Phys. Rev.* **A22** (1980), 1491.
10. Zaal G. J. *et al., Z. Phys.* **A290** (1979), 339.
11. Hühnermann H. *et al., Proceedings of the Sixth International Conference on Nuclei Far from Stability, Inst. Phys. Conf. Ser. No* 132 1992, p. 209.
12. Gangrsky Yu. P. *et al., Euro. Phys. J.* **D11** (2000), 341.
13. Ahmad S. A. *et al., Z. Phys.* **A321** (1985), 35.
14. Asaga T. *et al., Z. Phys.* **A359** (1997), 237.
15. Semmes P. and Ragnarsson I., The Particle + Rotor Model and Its Applications. unpublished, distributed at the Hands-on Nuclear Structure Theory Workshop, Risö, Denmark, June 15–26, 1992, and references therein.
16. Moskowitz P. A. and Lombardi M., *Phys. Lett.* **46B** (1973), 334.
17. Atomic Energy Levels, the Rare Earth Elements. Natl. Bur. Stand. Ref. Data Ser., Natl. Bur. Stand. (U.S.) Circ. No. 60, edited by W. C. Martin, R. Zalubas and L. Hagen (U.S. GPO, Washington, District of Columbia, 1978).
18. Pfeufer V. *Z., Phys.* **D4** (1987), 351.
19. Becker O. *et al., Phys. Rev.* **A48** (1993), 3546.
20. Enders K. *et al., Phys. Rev.* **A56** (1997), 265.
21. Enders K. *et al., Z. Phys.* **D42** (1997), 171.

Hyperfine Interactions (2005) 162:147–157
DOI 10.1007/s10751-005-9219-8

Laser Based Techniques for Ultra Trace Isotope Production, Spectroscopy and Detection

KLAUS D. A. WENDT[1,*], KLAUS BLAUM[1], CHRISTOPHER GEPPERT[1], PETER MÜLLER[1], WILFRIED NÖRTERSHÄUSER[1], ANNETTE SCHMITT[1], PHILIPP SCHUMANN[1], NORBERT TRAUTMANN[2] and BRUCE A. BUSHAW[3]
[1]*Institut für Physik, Johannes Gutenberg Universität, D-55099, Mainz, Germany*
e-mail: klaus.wendt@phys.uni-mainz.de
[2]*Institut für Kernchemie, Johannes Gutenberg Universität, D-55099, Mainz, Germany*
[3]*Pacific Northwest National Laboratory, Richland, WA 99352, USA*

Abstract. A variety of research activities in the field of fundamental and applied nuclear physics has evolved in the last years using resonantly tuned radiation from powerful lasers. The technique of resonance ionization spectroscopy has delivered outstanding results and found broad acceptance in the last years as a particularly efficient and highly selective method for rare and exotic radioisotope studies. It is used for production, spectroscopy and detection of these species and provides complete isobaric, high isotopic and even some isomeric selection, which altogether is needed for on-line investigation of short lived species far off stability as well as for ultra trace determination. Good overall efficiency pushes the experimental limits of detection in elemental trace analysis down to below 10^6 atoms per sample, and additionally isotopic selectivity as high as 3×10^{12} has been demonstrated. The widespread potential of resonance ionization techniques is discussed, focusing on the experimental arrangements for applications in selective on-line isotope production, spectroscopy of rare radioisotopes and ultra trace determination of radiotoxic isotopes like ^{238}Pu to ^{244}Pu, ^{135,137}Cs, ^{89,90}Sr or ^{41}Ca in environmental, technical and biomedical samples.

Key Words: laser spectroscopy, laser development, nuclear structure, radioisotopes, resonance ionization, ultra trace detection.

1. Introduction

As discussed extensively during this meeting on "Laser Methods In The Study of Nuclei, Atoms and Molecules," the atomic nucleus with its electromagnetic properties, determined by spin, nuclear moments and charge radius interacts significantly with the electronic shell. Thus atomic spectroscopy can reveal a detailed view into behavior, structure and stability of nuclear matter in its ground

---

* Author for correspondence.

state. Based on detailed studies along the valley of β-stability of the nuclear chart, the extension far out into the range of unstable short-lived radioactive nuclides are of major interest, e.g., for the development of corresponding nuclear models. Laser based techniques are ideal for this task, as they combine selectivity and sensitivity. In addition, they also have found widespread applications for all kind of analytics of rare isotopes. The early stages in this field of 'lasers in nuclear physics' are nicely resumed in the proceedings of corresponding conferences [1, 2]. Reviews on the topic of using atomic spectroscopy for studying nuclear physics have been published by E. W. Otten with the status of 1988 [3], by R. Neugart in 2002 [4] and by H. -J. Kluge and W. Nörtershäuser in 2003 [5].

Here we want to limit the presentation to the particular technique of laser resonance ionization, where lasers are used for efficient step-wise optical excitation of atoms by absorption of resonantly tuned light up to final ionization. The outstanding features, which result from this process, include selective and nearly background free production, investigation and detection of specific molecular, elemental and even isotopic species. They have attracted already early the interest of researchers from various fields. Apart from fundamental studies on the properties of intermediate and high-lying atomic and molecular states, many analytical applications of the technique were proposed and rapidly developed. A typical example is the sensitive trace detection of rare atomic or molecular species in technosphere and nature. Here particularly successful applications of resonance ionization have been realized [6, 7]. One of those is the determination of natural or anthropogenic ultra trace isotopes with relative abundances far below 10^{-9} for geological, cosmochemical or biomedical studies [8, 9]. In addition the selective ionization of artificially produced short-lived and exotic radioisotopes at on-line mass separator facilities for fundamental studies has become very popular due to striking advantages. Laser based experimental techniques applying resonant excitation developed for all these activities are based on very similar concepts and experimental components. In the field of on-line applications resonance ionization has to compete with conventional ionization techniques like surface, plasma or electron cyclotron resonance ionization processes. For the ultra trace analysis of rare, usually very long-lived radioisotopes the detection of their radioactive decay within a reasonable measuring time by radiometric techniques is either inefficient or non-specific. In that case direct counting of the number of atoms via selective ionization becomes favorable, in particular concerning long lived α- and β-emitters. Here dedicated mass spectrometric methods, like ICP-MS [10] and in particular AMS [11] are powerful competing techniques to laser ionization. Alternatively modern quantum optical approaches, like atomic traps, are just coming up with very promising specifications for future applications, concerning analytics [12] as well as basic fundamental studies [13].

Striking advantages of resonance ionization, which nominate the method for the different research fields discussed above, are

1. almost complete suppression of atomic or molecular isobaric interferences, obtained by the uniqueness of optical transitions, especially in multi-step optical excitation schemes;
2. good overall efficiency with ionization probability as high as 20% and corresponding analytical sensitivity and detection limits in the fg to ag range per sample, due to the large cross sections of optical excitation and ionization steps in the many Mb range;
3. excellent isotopic selectivity, achieved by high resolution techniques. Optical selectivity can be further increased by omitting Doppler broadening either in multi-step excitation processes or by application of an artificial Doppler-shift in collinear excitation on fast atomic or ionic beams.

Combination of resonant ionization with ion trapping and cooling techniques can further enhance these specifications by suppression of interferences from competing ionization processes, i.e., surface ionization, as well as by generating temporally well controlled ion beams of low emittance, which simplify subsequent detection or investigation [14].

2. Theoretical background of resonance ionization

Optical resonance excitation of an atomic species from the ground state up to final ionization can follow a number of different path ways. Various possible excitation ladders are compared in Figure 1. For an ionization potential of about 5 to 9 eV, which is typical for more than 80% of the elements, at least two, but usually three optical excitation steps in the ultraviolet ($\Delta E \approx 3.5$ eV and beyond) to visible ($\Delta E \approx 2$ eV) spectral range are used. Excitation along first and second steps of all channels indicated can be carried out very efficiently: Typical atomic lifetimes for permitted transitions between these low-lying states are of the order of a few 10 ns while the cross-section for absorption of a photon in resonance is $\sigma = \lambda^2/2\pi \approx 10^{-10}\,\mathrm{cm}^2$ for the idealized case of an atom at rest. As the excitation probability is given by $dW(t) = \sigma \cdot J(t) \cdot dt$, a flux of only about $J(t) \approx 10^{18}$ photons /(cm^2 • s) is sufficient for saturation of such a transition. Due to the different velocities of the moving atoms, in the real experiment an effective cross section for individual velocity classes has to be considered. This is lower by about a factor of up to 100, depending on the collimation of the atomic ensemble or beam. Nevertheless typical pulsed and continuous lasers are easily saturating. Direct non-resonant ionization into the continuum, as shown on the left hand side of Figure 1 is rather unfavorable due to the low cross section of only about 10^{-17} cm^2. Fortunately this 'bottle-neck' of the resonance ionization process can be avoided in almost all cases. Elements with a rich atomic

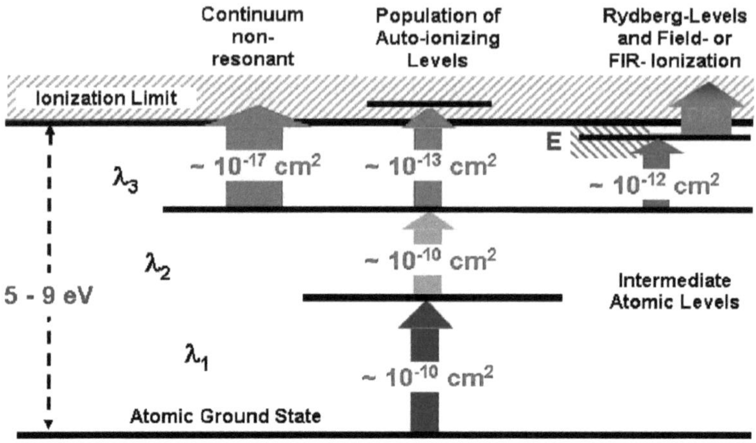

Figure 1. Comparison of different possible excitation schemes for resonance ionization. Typical cross-sections for the individual excitation and ionization steps are indicated.

spectrum, like e.g., lanthanides, actinides or many transition metals, exhibit auto-ionizing states, i.e., doubly excited states located above the first ionization potential. These decay with short life time of far less than 1 ns into a positively charged ion and an electron. Alternatively excitation into a high-lying Rydberg state and subsequent efficient field- or far infrared ionization can be used for all other elements. In both cases also the ionization step must be induced by a tunable laser set into resonance, which significantly increases efficiency and selectivity by reducing non-resonant background. In that case the ionization probability is easily increased by about two to three orders of magnitude up to saturation.

Usually powerful tunable pulsed laser systems are in use, which provide a well suited profile of spectral power density, easy handling and reliability and thus dominate the field of resonance ionization. Tunable continuous wave laser systems deliver high power density only in a very narrow spectral profile. Thus they are just applied for dedicated experimental arrangements, where high resolution and correspondingly high optical (isotopic) selectivity is required.

In the case of pulsed laser RIMS, a quantitative description of the excitation and ionization process can be worked out by simple rate equations, disregarding all coherent effects. The generalized rate equation for each participating level $k = 1,2,3,4$ of a multi-step excitation is given by the individual induced and spontaneous processes

$$\frac{d}{dt}N_k(t) = \sigma_{k-1} \cdot J_{k-1}(t) \cdot (N_{k-1}(t) - N_k(t)) - \sigma_k \cdot J_k(t) \cdot (N_k(t) - N_{k+1}(t))$$
$$-\Gamma_{k,k-1} \cdot N_k(t) + \Gamma_{k+1,k} \cdot N_{k+1}(t) - \gamma_k \cdot N_k(t)$$

with $N_k(t)$ being the level population, $J_k(t)$ the photon density of the laser field interconnecting level k with level $k + 1$, $\Gamma_{k,k-1}$ the spontaneous transition rate from level k to level $k-1$, given directly by the Einstein A-factor $\Gamma_{k,k-1} = A_{k,k-1}$ and γ_k comprising the spontaneous decay rates from level k into lower states, which are not further contributing, as well as the ionization process itself. If all the atomic parameters of the excitation and ionization scheme are known, the profile and intensity of the ionization signal can be determined with the use of the well known Lorentzian frequency dependence of the individual optical cross sections.

For a precise calculation the spatial and temporal laser profile, the spectral frequency distribution as well as the three-dimensional velocity distribution of the atomic beam must be properly taken into account. This leads to tedious numerical integrations and convolutions but delivers precise theoretical predictions.

This simple approach is just not valid for the case of multi-step excitation with narrow bandwidth continuous laser excitation, where the coherences between the atomic levels involved, induced by the different laser fields, must be properly considered. This can be done by applying the density matrix formalism, which leads to rather complex asymmetric line shapes in the multi-dimensional space of the detunings of the individual laser frequencies [15].

Obviously high efficiency in resonance ionization would involve either

(A) pulsed atomization well synchronized to the laser pulses, e.g., by pulsed release or desorption of atoms from a target or beam source,
(B) the use of ultimately high repetition rate lasers together with continuous atomization, or alternatively
(C) the combination of low repetition rate lasers with a buffer gas cell, as e.g., realized in an ion guide laser ion source [16].

The first process (A) is affected by unspecific background, i.e., ionized molecules and clusters resulting from the pulsed heating or desorption process. Thus, it has been dominantly applied to on-line studies on artificially produced radioisotopes of e.g., refractory elements, which are not easily accessible by other techniques, e.g., at experiments at ISOLDE/CERN [17–19].

In the second approach (B) the laser duty cycle directly affects the achievable overall efficiency and must be chosen in the high kHz range. Furthermore the location of resonance ionization is usually placed inside a widely closed hot cavity or transfer tube to form a so-called resonance-ionization-laser-ion-source (RILIS) [20–23]. The atoms have the chance to pass through the laser beam very often and in this way the overall ionization probability is strongly enhanced up to values of ~20%. Again background, stemming mainly from the unavoidable competing process of surface ionization on the hot cavity walls leads to significant interferences and limitations, which must be suppressed. For this task combinations of RILIS with repelling electrodes and the localization of the resonance ionization region inside a quadrupole ion guide without any wall contact are developed. Filled with buffer gas, this latter device can also serve for cooling and bunching of the resulting ion beam to enhance temporal and spatial beam quality [14].

In approach (C) the atoms are stored in a noble gas buffer gas cell for many ms with the interaction volume being illuminated by laser light. Laser repetition rates of ~100 Hz are sufficient for efficient resonant ionization. The ions are guided to a nozzle by electric fields, flushed out by a gas jet, separated from the buffer gas by skimmers, and transferred to subsequent experiments or detection. The overall efficiency here is about 10^{-4}. Very recently this arrangement has enabled the first spectroscopic studies on fermium with about 10^{10} atoms of ^{255}Fm [24].

3. Pulsed laser systems for resonance ionization

The discussion given here shall be limited to high repetition rate laser systems as being typically used for resonance ionization: Until very recently these consisted of two to three tunable dye lasers pumped by at least two powerful Cu-vapor lasers (Cvl) [25]. Outstanding properties are the high repetition rate of 6 to 11 kHz and the broad tuning range through the use of a number of suitable dyes as well as frequency doubling units, together with a well adapted short pulse length of ~20 ns and spectral width of a few GHz. However, drawbacks of such a system are its size, high maintenance efforts and costs of the pump lasers. This has led to the development of novel, powerful and easy-to-handle solid state laser systems, which are presently conquering the field [26]. Nowadays highly reliable Q-switched and intracavity doubled Nd:YAG lasers with repetition rates of 1–25 kHz, power of 50 to 100 W at 532 nm and pulse lengths in the 100 ns regime are commercially available. These are well suited for simultaneously pumping of up to three titanium–sapphire (Ti:Sa) lasers as well as further solid state amplifier stages. Each solid state laser provides an average power of up to 3 W in a tuning range from ~700 to ~1,000 nm, line width around 2–5 GHz and pulse duration of ~60 ns. For resonance ionization the three lasers must be synchronized, which is achieved with Pockels cells used as intracavity Q-switches. Single pass frequency doubling and tripling is possible and required to cover the blue and ultraviolet spectral ranges.

As compiled in Figure 2, similar to dye lasers, Ti:Sa laser systems are nowadays generally capable of resonantly ionizing the far majority of elements in the periodic table. While the Cvl pumped dye lasers already have proven their performance successfully on more than 20 elements for analytical and on-line applications, excitation schemes on many remaining elements, in particular concerning application of the rather novel Ti:Sa laser system must still be searched and investigated spectroscopically in the near future. These activities are in progress at a number of on-line facilities and laboratories worldwide [14].

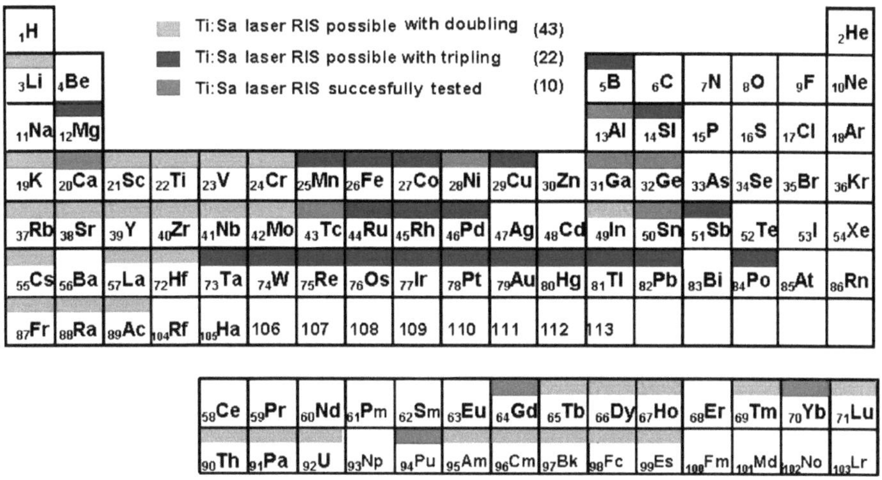

Figure 2. Periodic system of the elements, the *shaded boxes* indicate the accessibility of the individual element to resonance ionization using Ti:Sa lasers.

In the frame of analytical applications on elemental species, pulsed laser RIMS is frequently used for trace and ultra trace determination on long-lived radiotoxic isotopes of actinides and technetium, where necessary suppression of isobars excludes conventional mass spectrometric techniques. In combination with thermal evaporation and simple and reliable time-of-flight mass spectrometers pulsed laser resonant ionization significantly surpasses the specifications of radiometric counting techniques in respect to detection limit and short measuring time by far [27].

4. Resonant processes using narrow bandwidth continuous wave lasers

Resonance ionization using continuous wave lasers provides high isotopic selectivity through utilization of the isotopic shift in high resolution spectroscopy. In cases where naturally occurring isotope shifts are small and insufficient to reach the required selectivity collinear geometry on fast atomic beams can be applied by utilizing the mass dependent Doppler shift. With kinetic energies in the range of several 10 keV, a large artificial isotope shift in the many GHz range is introduced, which strongly enhances isotopic selectivity and even can serve for tuning fixed frequency lasers into resonance to atomic transitions within the rest frame of the fast moving atom. This strongly simplifies the laser equipment needed. Furthermore, the conservation of the longitudinal energy spread during acceleration leads to a significant reduction in the velocity spread dependent Doppler width. The production of a fast atomic beam involves extensive machinery: i.e., ionization in a conventional ion source, acceleration of the ions to the necessary kinetic energy and subsequent efficient neutralization in a charge

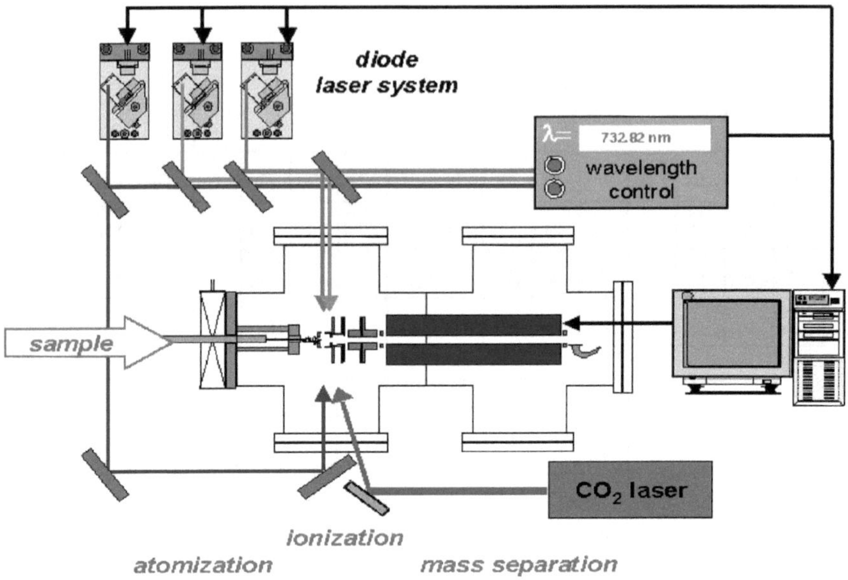

Figure 3. Schematic sketch of the compact experimental set-up for diode laser based high resolution RIMS at a quadrupole mass spectrometer for analytical applications requiring highest isotopic selectivity.

exchange medium. Fortunately these components are at least partly realized within the components of a mass separator, which makes this technology ideally suited for direct on-line use at exotic radioisotope production facilities of the ISOL type. The charge exchange process furthermore provides significant population of up to 10% in meta-stable atomic states, from which an efficient resonance excitation into high lying states close to the ionization limit is possible in a single excitation step.

Collinear RIMS was first applied for the determination of the ultra trace noble gas isotope ^{3}He in environmental samples [28], and later on at CERN for the investigation of nuclear charge radii and nuclear moments within the isotopic chain of Yb [29]. This technique was also used for a fast and sensitive laser based determination of ^{89,90}Sr in various environmental and technical samples [30].

A typical set-up for high resolution RIMS on a collimated thermal atomic beam is sketched in Figure 3. Multi-step excitation with narrow-band cw lasers is applied in perpendicular geometry in combination with a quadrupole mass spectrometer (QMS) or alternatively a magnetic sector field [31]. Due to its limited experimental expenditure and its compactness, this technique is especially well suited for trace analysis if compact and cheap diode lasers are used. Further advantages compared to other RIMS versions are the extremely high elemental and isotopic selectivity together with a good overall efficiency. This technique has already been applied successfully for ultra trace analysis of ^{90}Sr [32], ^{135,137}Cs [33] and ^{210}Pb [34]. In all applications isotope-selective stepwise excitation in an atomic beam delivered extremely high isotopic selectivity of up to 3×10^{12}, which has been demonstrated for the case of ^{41}Ca [35]. A recent review of these techniques in comparison to AMS and atomic trap trace analysis is found in [36].

5. Conclusions

The spectroscopic technique of resonance ionization on atomic species is nowadays realized in various experimental arrangements, which are specifically tailored to meet the demands of selective ionization at on-line facilities or those of ultra trace determination of rare radioisotopes in various samples for different applications. Laser systems, mass spectrometric components and ion detectors are well adapted both for the on-line and shift work as well as for the routine-like operation in dedicated environmental or biomedical analytical studies. In analytical applications the technique is complementary in a number of aspects to AMS and extends the capabilities of conventional mass spectrometry. Outstanding features are the isobaric and, if required, additional high isotopic selectivity, high ionization efficiency and the corresponding low detection limits. Ultimate specifications reported include isobaric suppression of numerous orders of magnitude, depending on the species under study, isotope selectivities of up to 3×10^{12}, overall resonance ionization efficiencies of 20% and beyond and a precision in isotope ratio measurements of ~1%. In analytic applications detection limits below 10^6 atoms are routinely realized, e.g., for Pu isotopes. The method is further improved by the ongoing development on the side of the lasers. It is currently extended to cover the far majority of elements within the periodic table for the purpose to further enhance the knowledge on rare and exotic species.

References

1. Bemis C. E. Jr. and Carter H. J. (eds.), *Lasers in Nuclear Physics*, Harwood Academic Publ., New York, 1982.
2. Application of Lasers in Atomic Nuclear research, *Proc. V. Int. Workshop*, Poznan, Dubna JINR Press, 2001.
3. Otten E. W., Nuclear radii and moments of unstable isotopes, In: D.A. Bromley, *Treatise on Heavy-Ion Science*, Vol. 8, Plenum Press, London, 1989, p. 517–638.
4. Neugart R., Lasers in nuclear physics, *Eur. Phys. J.* **A15** (2002), 35.
5. Kluge H.-J. and Nörtershäuser W., Laser for nuclear physics, *Spectrochim. Acta* **B58** (2003), 1031.
6. Lethokov V. S. (ed.), *Laser Photoionization Spectroscopy*, Academic Press, Orlando, 1987.
7. Hurst G. S. and Payne M. G. (eds.), *Principles and Applications of Resonance Ionisation Spectroscopy*, Adam Hilger Publ., Bristol, 1988.
8. Payne G., Deng L. and Thonnard N., *Rev. Sci. Instrum.* **65** (1994), 2433.
9. Wendt K., *Eur. Mass Spectrom.* **8** (2002), 273.

10. Montaser A. (ed.), *Inductively Coupled Plasma Mass Spectrometry*, Wiley, John & Sons, New York, 1998.
11. Tuniz C., Bird J. R., Fink D. and Herzog G. F. (eds.), *Accelerator Mass Spectrometry*, CRC Press LLC, Boca Raton, 1998.
12. Chen C. Y., Li Y. M., Bailey K., O'Connor T. P., Young L. and Lu Z.-T., *Science* **286** (1999), 1193.
13. Sprouse G. D. and Orozco L. A., *Annu. Rev. Nucl. Part. Sci.* **47** (1997), 429.
14. Wies K. *et al.*, contribution to this issue.
15. Nörtershäuser W., Bushaw B. A., Müller P. and Wendt K., *Appl. Opt.* **39** (2000), 5590.
16. Kudryavtsev Y., Andrzejewski J., Bijnens N., Franchoo S., Gentens J., Huyse M., Piechaczek A., Szerypo J., Reusen I., Van Duppen P., Van Den Bergh P., Vermeeren L., Wauters J. and Wöhr A., *Nucl. Instrum. Methods Phys. Res.* **B114** (1996), 350.
17. Krönert U., Becker S., Hilberath T., Kluge H.-J. and Schulz Ch., *Appl. Phys.* **A44** (1987), 339.
18. Krönert U., Becker S., Bollen G., Gerber M., Hilberath T., Kluge H.-J. and Passler G. *Nucl. Instrum. Methods Phys. Res.* **A300** (1991), 522.
19. Sauvage J., Boos N., Cabaret L., Crawford J. E., Duong H. T., Genevey J., Girod M., Huber G., Ibrahim F., Krieg M., Le Blanc F., Lee J. K. P., Libert J., Lunney D., Obert J., Oms J., Peru S., Pinard J., Putaux J. C., Roussiere B., Sebastian V., Verney D., Zemlyanoi S., Arianer J., Barre N., Ducourtieux M., Forkel-Wirth D., Le Scornet G., Lettry J., Richard-Serre C. and Veron C., *Hyp. Int.* **129** (2000), 303.
20. Andreev S. V., Mishin V. I. and Letokhov V. S., *Opt. Commun.* **57** (1986), 317.
21. Ames F., Brumm T., Jäger K., Kluge H.-J., Suri B. M., Rimke H., Trautmann N. and Kirchner R., *Appl. Phys. B* **51** (1990), 200.
22. Mishin V. I., Fedoseyev V. N., Kluge H.-J., Letokhov V. S., Ravn H. L., Scheerer F., Shirakabe Y., Sundell S. and Tengblad O., *Nucl. Instrum. Methods Phys. Res. B* **73** (1993), 550.
23. Koester U., Fedoseyev V. N. and Mishin V. I., *Spectrochim. Acta. B* **58** (2003), 1047.
24. Sewtz M., Backe H., Dretzke A., Kube G., Lauth W., Schwamb P., Eberhardt K., Grüning C., Thörle P., Trautmann N., Kunz P., Lassen J., Passler G., Dong C. Z., Fritsche S. and Haire R. G. *Phys. Rev. Lett.* **90** (2003), 163002.
25. Fedoseyev V. N., Koester U., Weisman L., Mishin V. I., Horn R., Huber G., Lassen J., Wendt K., Fedorov D. V. and the ISOLDE Collaboration, *Nucl. Instrum. Methods Phys. Res.* **204 C** (2003), 353.
26. Grüning C., Huber G., Klopp P., Kratz J. V., Kunz P., Passler G., Trautmann N., Waldek A. and Wendt K., *Int. J. Mass Spectrom.* **235** (2004), 171.
27. Passler G., Erdmann N., Hasse H. U., Herrmann G., Huber G., Köhler S., Kratz J. V., Mansel A., Nunnemann M., Trautmann N. and Waldek A., *Kerntechnik* **62** (1997), 85.
28. Kudryavtsev Y. A., Petrunin V. V., Sitkin V. M. and Lethokov V. S., *Appl. Phys. B* **48** (1989), 93.
29. Schulz C., Arnold E., Borchers W., Neu W., Neugart R., Neuroth M., Otten E. W., Scherf M., Wendt K., Lievens P., Kudryavtsev Y. A., Lethokov V. S., Mishin V. I. and Petrunin V. V., *J. Phys. B* **24** (1991), 4831.
30. Wendt K., Bhomwick G. K., Bushaw B. A., Herrmann G., Kratz J. V., Lantzsch J., Müller P., Nörtershäuser W., Otten E.-W., Schwalbach R., Seibert U. A., Trautmann N. and Waldek A., *Radiochim. Acta* **79** (1997), 183.
31. Bushaw B. A., *Prog. Anal. Spectrosc.* **12** (1989), 247.
32. Bushaw B. A. and Cannon B. D., *Spectrochim. Acta* **52B** (1997), 1839.

33. Pibida L., Nörtershäuser W., Hutchinson J. M. R. and Bushaw B. A., *Radiochim. Acta*, **89** (2001), 161.
34. Bushaw B. A., RIS-92, *Inst. Phys. Conf. Series* **128** (1992), 31.
35. Müller P., Bushaw B. A., Blaum K., Diel S., Geppert Ch., Nähler A., Trautmann N. and Wendt K., *Fresenius J. Anal. Chem.* **370** (2001), 508.
36. Lu Z. T. and Wendt K. D. A., *Rev. Sci. Instrum.* **74** (2003), 1169.

Hyperfine Interactions (2005) 162:159–166
DOI 10.1007/s10751-005-9218-9

Resonance Ionization Mass Spectrometry (RIMS) with Pulsed and CW-Lasers on Plutonium

P. KUNZ[1,*], G. HUBER[1], G. PASSLER[1], N. TRAUTMANN[2] and K. WENDT[1]
[1]*Institut für Physik, Universität Mainz, D-55099 Mainz, Germany;*
e-mail: pkunz@uni-mainz.de
[2]*Institut für Kernchemie Universität Mainz, D-55099 Mainz, Germany*

Abstract. The detection of long-lived plutonium isotopes in ultra-trace amounts by resonance ionization mass spectrometry (RIMS) is a well-established routine method. Detection limits of 10^6 to 10^7 atoms and precise measurements of the isotopic composition have been achieved. In this work multi-step resonance ionization of plutonium atoms has been performed with tunable lasers having very different output intensities and spectral properties. In order to compare different ways for the resonance ionization of plutonium broadband pulsed dye and titanium:sapphire lasers as well as narrow-band cw-diode and titanium:sapphire lasers have been applied for a number of efficient excitation schemes. It has been shown, that for identical excitation schemes the optical isotope selectivity can be improved by using cw-lasers (bandwidths < 10 MHz) instead of pulsed lasers (bandwidths > 2 GHz). Pulsed and cw-laser systems have been used simultaneously for resonance ionization enabling direct comparisons of pulsed and continuous ionization processes. So far, a three-step, three-color laser excitation scheme has been proven to be most practical in terms of efficiency, selectivity and laser wavelengths. Alternatively a newly discovered three-step, two-color excitation scheme which includes a strong two-photon transition from an excited state into a high-lying autoionizing state yields similar ionization efficiencies. This two-photon transition was characterized with respect to saturation behavior and line width.

Key Words: plutonium, resonance ionization mass spectrometry, two-photon transitions, ultra-trace analysis.

1. Isotope selective ultra-trace analysis of plutonium

The most abundant plutonium isotopes are produced in nuclear reactors via a series of neutron capture reactions starting with ^{238}U:

$$^{238}\mathrm{U}(n,\gamma)^{239}\mathrm{U} \xrightarrow{\beta^-} {}^{239}\mathrm{Np} \xrightarrow{\beta^-} {}^{239}\mathrm{Pu}$$

$$^{239}\mathrm{Pu}(n,\gamma)^{240}\mathrm{Pu}(n,\gamma)^{241}\mathrm{Pu}(n,\gamma)^{242}\mathrm{Pu}(n,\gamma)^{243}\mathrm{Pu}(n,\gamma)^{244}\mathrm{Pu}$$

Due to the large scale utilization of nuclear fission since more than 60 years for civil as well as military purposes significant amounts of radio nuclides have been

* Author for correspondence.

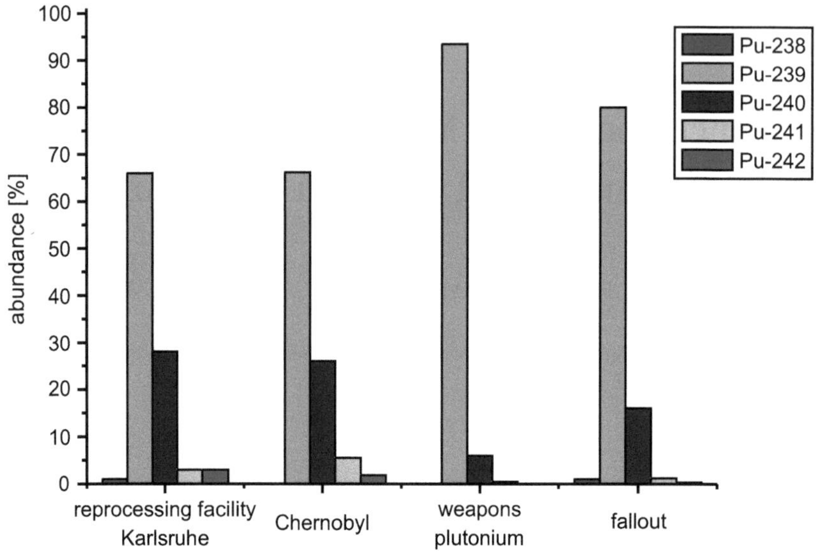

Figure 1. Typical isotope compositions for different origins or forms of utilization of plutonium [2–4].

produced, which would naturally occur only in very low concentrations or not at all. Especially long-lived radionuclides – such as several plutonium isotopes – can contaminate the environment for a long time and can pose serious health risks because of their high radiotoxicity [1]. Plutonium is released into the environment mainly through nuclear weapons tests, accidents in nuclear facilities, satellite or plane crashes and ship or submarine accidents as well as controlled emissions from nuclear facilities [2].

Because of its high toxicity and its important and security-relevant applications (e.g., energy production, nuclear weapons) the monitoring of the environment, the surveillance of people and facilities is a major issue. Therefore suitable analytical methods are required. Resonance ionization mass spectrometry enables detection limits of 10^6 to 10^7 atoms of plutonium per sample [4]. The very high element selectivity as a result of the resonant laser ionization process practically eliminates all isobaric interference which is a huge problem for other mass spectrometric methods with less selective ionization procedures. Furthermore, it is possible to determine accurate isotope ratios independent of the half-life or the decay mode of the respective Pu isotope [5]. The isotope composition yields information on the origin or the application of the plutonium in question [2–4] (see Figure 1).

2. Resonance ionization mass spectrometry on plutonium – experimental setup

The first step in the measurement of a plutonium sample via RIMS is a chemical separation process and the preparation of a sample for an efficient atomic beam source, which is combined with a suitable laser system for the resonance ionization

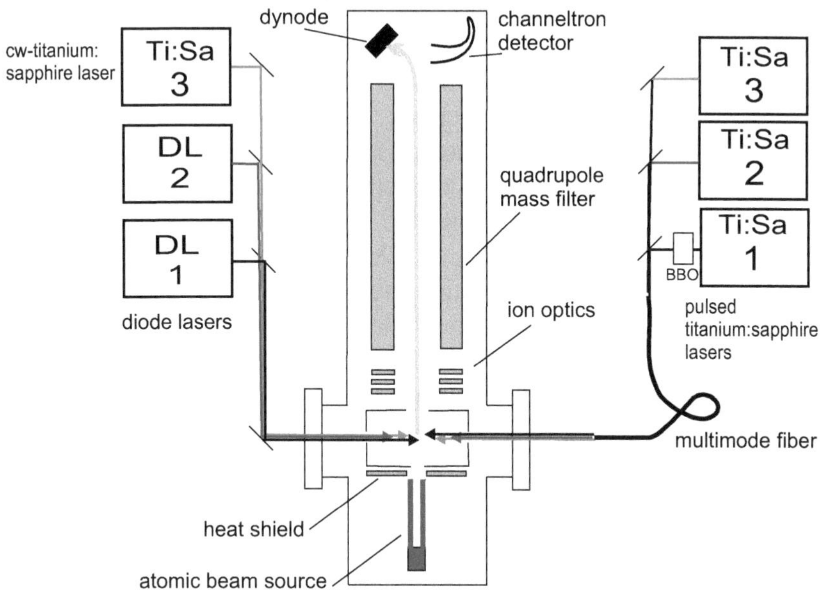

Figure 2. Schematics of a compact RIMS-apparatus optimized for narrow-band cw-lasers (tunable diode and cw-titanium:sapphire lasers). The use of a broadband pulsed laser system is optional.

and a mass spectrometer. In order to achieve a high overall efficiency the properties of these components must be optimized [6]. The multi-step resonance ionization of the atomic beam can be performed with tunable lasers having varying output intensities and spectral properties. A newly developed compact RIMS apparatus (Figure 2) was used to examine the resonance ionization of plutonium with narrow-band cw-lasers for the first time. Its main features are a collimated atomic beam source and a continuously operating quadrupole mass spectrometer [6]. The laser system consists of tunable diode and titanium:sapphire lasers with line widths <10 MHz. Broadband pulsed laser systems consisting of tunable titanium:sapphire and dye lasers are also available. They operate at a repetition rate of 6.6–7 kHz with typical bandwidths of 2–3.3 GHz. By using both laser systems simultaneously it was possible to perform comparative measurements between well-understood pulsed and cw-laser excitation.

Another RIMS apparatus optimized for pulsed ionization is routinely used for analytical measurements of Pu-isotope ratios in ultra-trace amounts (femtogram per gram) but also for the determination of ionization potentials of actinides with very small samples of less than 10^{12} atoms. It is based on a broadband pulsed laser system and a time-of-flight mass spectrometer [7, 8].

3. RIMS on plutonium with narrow-band cw-lasers and broadband pulsed lasers

The spectroscopic measurements on plutonium have been performed with very small sample sizes. Less than 10^{14} atoms were used for all the experiments, which are discussed extensively in [6]. They include the determination of line widths, saturation intensities of transitions between bound as well as transitions into autoionizing states, the examination of hyperfine structures of ^{239}Pu in a three-step, three-color excitation scheme, the measurement of isotope shifts for the most stable Pu-isotopes and isotope ratio measurements. Some of the most interesting results are presented in the following paragraphs.

Line widths: The experimental line forms of the three-step, three-color excitation scheme $\lambda_1 = 420.77$ nm, $\lambda_2 = 808.29$ nm, $\lambda_3 = 790.28$ nm (Figure 5b) were determined for each excitation step by first optimizing the RIMS ion signal and then scanning one laser while the other two lasers remained fixed. They are a result of the convolution of the laser bandwidths, the Doppler broadening in the atomic beam, the natural line width of the transition and power broadening in the case of saturation. With narrow-band cw-lasers the bound state transitions are mainly determined by power and Doppler broadening. The overall line widths of the transitions into autoionizing states are mostly dominated by their natural line widths. In contrast to the broadband pulsed excitation the bandwidths of cw-lasers do not contribute significantly to the overall line form.

For the first and the second excitation step the cw-saturation powers were determined for a beam size of approximately 0.015 cm^2 by measuring the ion signal intensity for different laser powers: $P_1^{sat} = 0.050(7)$ mW, $P_2^{sat} = 10(2)$ mW. The third excitation step could not be saturated with the available laser power. Therefore only a lower limit of $P_3^{sat} > 500$ mW was estimated here. Due to the very small sample sizes and the resulting low ion count rates the highest available laser powers ($P_1^{cw} = 5$ mW, $P_2^{cw} = 17$ mW, $P_3^{cw} = 600$ mW) were used in the experiments with cw-lasers leading to the results outlined below. The first excitation step λ_1 could be saturated approximately one hundred times and the second step λ_2 one to two times. The measured line width of $\nu_2 = 125(4)$ MHz has a Voigt-profile with a dominating Gaussian component. This suggests that the line width is mostly determined by the Doppler broadening of the atomic beam. In the line shape of the well saturated step λ_1 with $\nu_1 = 247(10)$ MHz the most significant component is a Lorentzian. The line shape of transition λ_3 can be approximated very well by a Lorentzian even though it could not be saturated. The measured line width $\nu_3 = 3.4(1)$ GHz is large in comparison to the Doppler broadening. It is determined by the natural line width of the autoionizing state (see Figure 3a). Several autoionizing states of plutonium had been examined with narrow-band cw-lasers and all of them have line widths larger than 1.7 GHz [6].

Isotope shift: For the determination of isotope ratios with RIMS it is essential to know the isotope shifts in each excitation step, because the lasers have to be tuned to the different isotopes during the measurement. The required accuracy of the tuning process depends strongly on the bandwidths of the excitation lasers. Figure 3 shows the isotope shifts between ^{242}Pu and ^{244}Pu in a three-step, three-

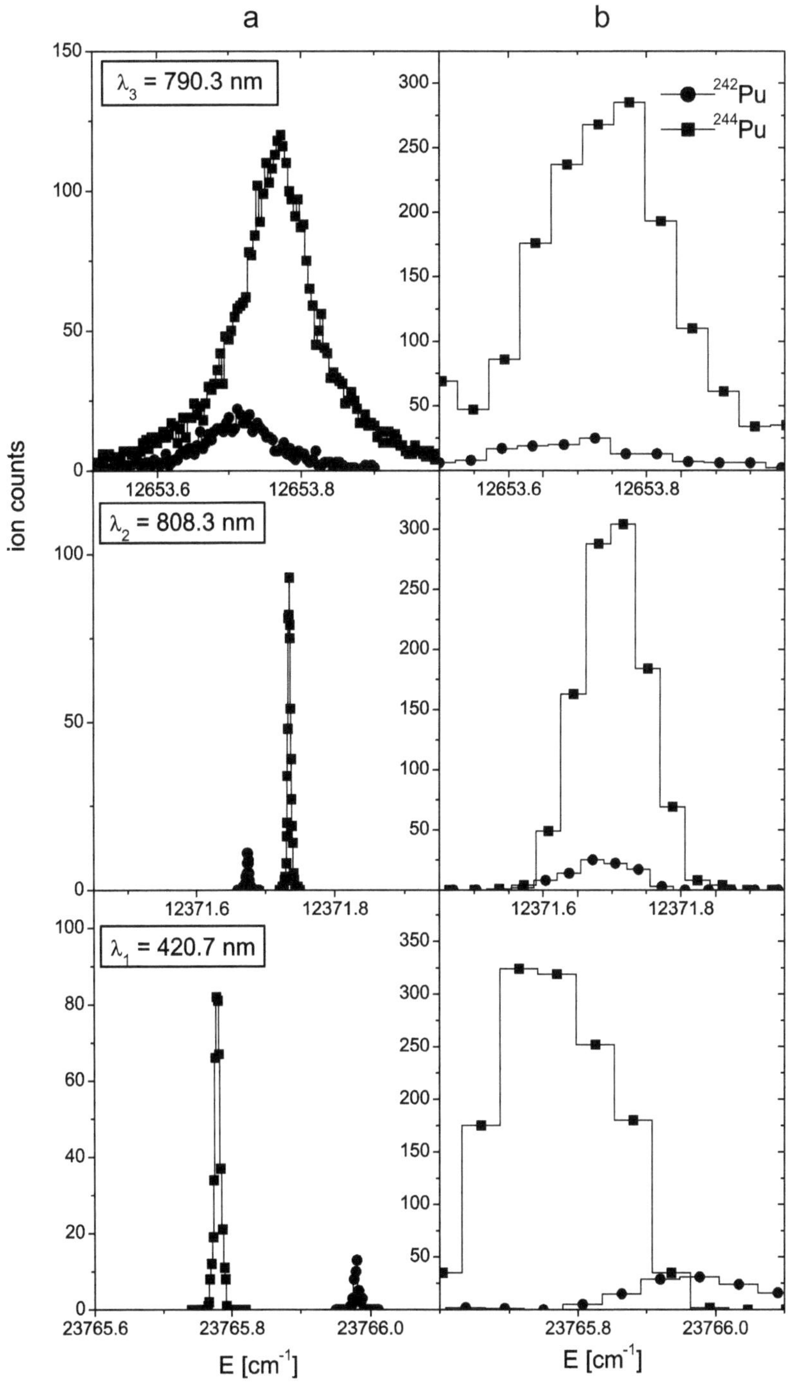

Figure 3. Isotope shift between ^{242}Pu and ^{244}Pu in a three-step, three-color excitation scheme measured with narrow-band cw-lasers (a) and broadband pulsed lasers (b). The ^{242}Pu:^{244}Pu isotope ratio in the sample used for these measurements was approx. 1:10.

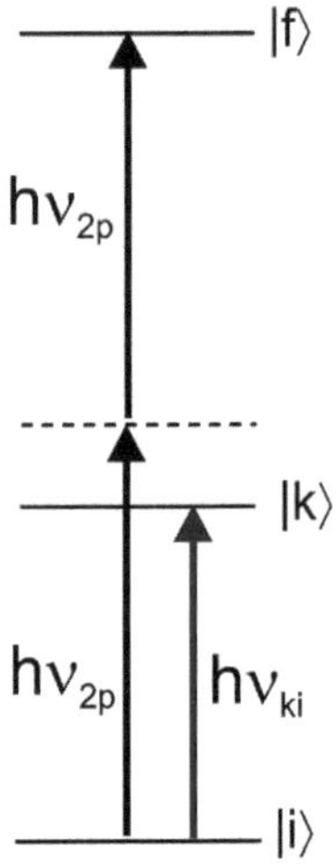

$$W_{fi}(\nu_{if} = 2\nu_{2p}) = \frac{5.55}{\Gamma_{fi}} \left|A_{if}\right|^2 I_{2p}^2 \tag{1}$$

$$A_{if} = \sum_k \frac{\langle f|z|k\rangle\langle k|z|i\rangle}{h(\nu_{ki} - \nu_{2p})} \tag{2}$$

Figure 4. Schematic view of a two-photon transition $|i\rangle \rightarrow |f\rangle$ with the energy $2h\nu_{2p}$. The overall transition rate (1) is enhanced by intermediate states $|k\rangle$ expressed in a composite transition matrix element (2).

color excitation scheme measured with narrow-band cw-lasers (a) and broadband pulsed lasers (b). By using narrow-band cw-lasers a much higher resolution has been achieved.

Ionization efficiency: In relative efficiency measurements between pulsed and cw-RIMS an atomic plutonium beam was overlapped alternately with pulsed and cw-lasers, while other experimental parameters remained unchanged. They showed, that even with duty-cycle losses pulsed RIMS is three to four times more efficient than cw-RIMS [6]. This can be attributed to a better spectral and spatial overlap with the atomic beam for pulsed lasers and the fact that for cw-lasers the ionization step could not be saturated.

4. Three-step, two-color excitation of plutonium with broadband pulsed lasers

In an extensive study of resonance structures above the first ionization limit of plutonium with broadband pulsed lasers several strong two-photon transitions into autoionizing states were discovered [6]. A three-step, two-color excitation scheme, which includes such a two-photon transition ($\lambda_1 = 420.77$ nm, $\lambda_{2p} = 582.27$ nm, Figure 5a) was examined in detail. The first excitation step leads to an $5f^6 7s7p\ ^7D_1^o$ excited state which is the starting point for a two-photon transition into an autoionizing state 58113.86 cm^{-1} above the ground state $5f^6 7s^2\ ^7F_0$ [9].

The rate of a two-photon transition W_{fi} depends strongly on the laser intensity I_{2p} and the composite transition matrix element A_{if} (see Figure 4). Its total value

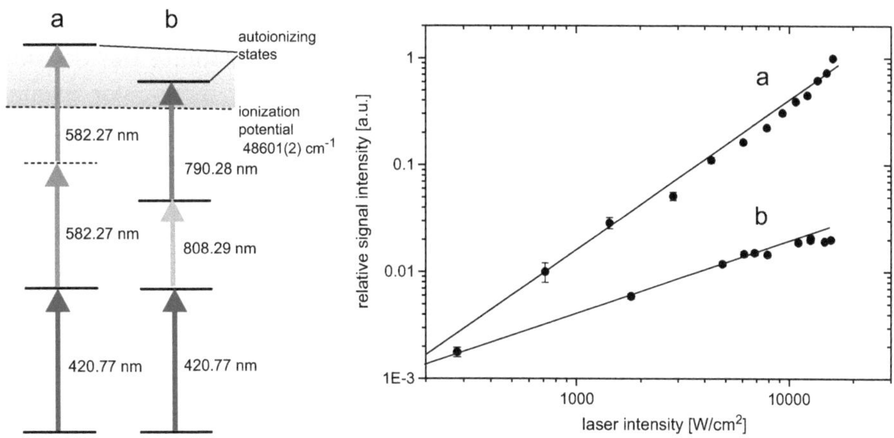

Figure 5. (a) Two-color, three-step excitation scheme with a two-photon transition into an autoionizing state. (b) Three-color, three-step excitation scheme with a one-photon transition into an autoionizing state. The rate for the two-photon transition (a) exhibits a nonlinear dependency on the laser intensity while the one-photon transition (b) shows a linear behavior as long as saturation effects are not dominant.

is significantly influenced by intermediate states $|k\rangle$ in the atomic level system. $h\nu_{2p}$ is the photon energy and Γ_{fi} the natural line width of the two-photon transition $|i\rangle \rightarrow |f\rangle$ with the frequency ν_{if} [10].

Under the condition that bound state transitions can be saturated with the available laser fluence, for a two-photon transition the frequency gaps between the two-photon resonance energy $h\nu_{2p}$ and the positions of contributing intermediate states $h\nu_{ki}$ have to be substantially larger than the two-photon excitation line width: $|\nu_{ki} - \nu_{2p}| \gg \Delta\nu_{2p}$ [11]. Then – according to equation (1) [10] in Figure 4–the transition rate W_{fi} depends on the laser intensity I_{2p} squared whereas a one-photon transition exhibits a linear dependence. In Figure 5 the experimental comparison between two-photon transition (a) and a one-photon transition (b) into an autoionizing state of plutonium is shown.

The laser bandwidth ($\approx$3.3 GHz) was larger than the Doppler broadening of the atomic beam (<1 GHz). Thus, the measured line width of the two-photon transition of 7.7(3) GHz is the result of a convolution of the Doppler broadening, the laser band-width and the natural line width of the autoionizing state. The ratio of the signal intensity to the non-resonant background (resonance enhancement factor) is almost one order of magnitude higher than the ratio usually observed in three-step, three-color excitation schemes (e.g., Figure 5b), where non-resonant ionization with photons of the first excitation step occurs from a high-lying excited bound state. This is not possible in the two-color excitation scheme (Figure 5a). The application of the two-color scheme yields similar overall efficiencies ($> 10^{-5}$) [9] like the use of the three-color scheme with overall efficiencies of $4 \cdot 10^{-5}$, but only two tunable lasers are required.

5. Conclusion

A new RIMS apparatus for the ultra-trace analysis of plutonium was developed. With this setup it was possible to examine three-step, three-color excitation schemes in plutonium for the first time with narrow-band cw-lasers and to compare the results directly with broadband pulsed laser excitation. Among other properties line widths, isotope shifts, hyperfine structures were analyzed and optical selectivity and efficiency measurements were performed. Furthermore, a new three-step, two-color excitation scheme featuring a strong two-photon transition into an autoionizing state was examined with pulsed lasers. It simplifies the excitation process by reducing the number of required tunable lasers, while it preserves the benefits of multi-step resonance ionization in terms of efficiency and selectivity.

Acknowledgements

The authors would like to thank the *Bundesministerium für Bildung und Forschung* (06MZ962I) and the *Stiftung Rheinland-Pfalz für Innovation* for financial support.

References

1. Taylor D. M., *Appl. Radiat. Isotopes* **46**(11) (1995), 1245-1252.
2. International Atomic Energy Agency, *Safe Handling and Storage of Plutonium, Safety Reports Series 9*, IAEA, Vienna, (1998).
3. Fetter S., Frolov V. A., Prilutsky O. F. and Sagadeev R. Z., *Sci. Glob. Secur.* **1** (1990), 225–302.
4. Nunnemann M., Erdmann N., Hasse H.-U., Huber G., Kratz J. V., Kunz P., Mansel A., Passler G., Stetzer O., Trautmann N. and Waldek A., *J. Alloys Compd.* **45** (1998), 271–273.
5. Passler G., Erdmann N., Hasse H. U., Herrmann G., Huber G., Köhler S., Kratz J. V., Mansel A., Nunnemann M., Trautmann N. and Waldek A., *Kerntechnik* **62** (1997), 85–90.
6. Kunz, P., Dissertation, Institut für Physik, Universität Mainz, (2004).
7. Grüning C., Huber G., Klopp P., Kratz J. V., Kunz P., Passler G., Trautmann N., Waldek A. and Wendt K., *Int. J. Mass Spectrom.* **235**(2) (2004), 171–178.
8. Erdmann N., Nunnemann M., Eberhardt K., Herrmann G., Huber G., Köhler S., Kratz J. V., Passler G., Peterson J. R., Trautmann N. and Waldek A., *J. Alloys Compd.* **271–273** (1998), 837–840.
9. Kunz P., Huber G., Passler G. and Trautmann N., *Eur. Phys. J., D* **29** (2004), 183–188.
10. Bloembergen N. and Levenson M. D., *Topics in Applied Physics*, Vol. 13, Springer, Berlin Heidelberg New York, 1976, p. 315.
11. Letokhov V. S., *Laser Photoionization Spectroscopy*, Academic Press, Inc., Orlando, 1987, p. 35.

Hyperfine Interactions (2005) 162:167–172
DOI 10.1007/s10751-005-9221-1

Atom Trap Trace Analysis of Ca Isotopes

S. HOEKSTRA*,†, A. K. MOLLEMA, R. MORGENSTERN, L. WILLMANN, H. W. WILSCHUT and R. HOEKSTRA
Atomic Physics, KVI, Rijksuniversiteit Groningen, Zernikelaan 25, 9747 AA, Groningen, The Netherlands;
e-mail: hoekstra@fhi-berlin.mgp.de

Abstract. In our experiment we aim at the detection of the rarest, naturally occuring calcium isotope ^{41}Ca by means of atom trap trace analysis. On basis of single-atom detection of ^{46}Ca our present sensitivity for ^{41}Ca is estimated to be 1 atom per hour at an abundance of 10^{-12}. To reach a sensitivity at the level of natural abundance, which is 10^{-14}, we need to reduce atomic beam losses. To achieve this, optical compression of the atomic beam is a promising option. We use Monte Carlo Simulations to demonstrate that optical compression of the atomic beam increases throughput of the atomic beam as well as isotope selectivity.

Key Words: atomic beams, Ca-41, isotope selection, magneto-optical trap, single atoms, trace analysis.

1. Introduction

Trace analysis of long-lived isotopes has become an important tool in modern science [13]. For the detection of trace elements Atom Trap Trace Analysis (ATTA) is a promising new technique [2, 4], next to established methods as accelerator mass spectrometry [11], low level counting [5] and resonance ionization mass spectrometry [7, 21].

In ATTA experiments, isotope selection is achieved by the repeated excitation (by a laser) of an optically accessible electronic transition in the neutral atom. Because of the isotope shift the scattering force induced by light of a fixed frequency is different for the different isotopes. The selectivity that can be obtained is determined by the ratio between the natural line width and the isotope shift of the pumping transition. In general isotope shifts are much larger than the natural line widths. The problem of Doppler broadening is circumvented by using atoms that are laser cooled and trapped in a magneto-optical trap (MOT) [17].

Recently ATTA was successfully used to date one million year old groundwater from the Nubian Aquifer (Egypt), by detecting the very low abundance of ^{81}Kr

* Author for correspondence.

† Present address: Fritz-Haber Institut der Max-Planck Gesellschaft, Faradayweg 4-6, 14195, Berlin, Germany.

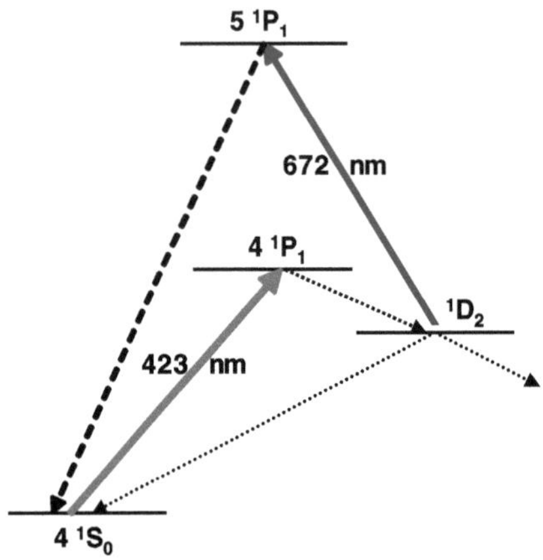

Figure 1. Relevant levels of calcium. Indicated are wavelengths and decay channels including the weak loss channel from $^1D_2$to the triplet system.

[19]. ATTA holds the potential to detect the long-lived isotope ^{41}Ca at its natural abundance level of 10^{-14} [9, 18]. This would open the possibility to perform radio-calcium dating with ^{41}Ca, which has a half-life of 10^5 years [9, 20]. Furthermore, ^{41}Ca can be used as a tracer to monitor osteoporosis on a molecular level [6]. It has been demonstrated that ATTA is already sufficiently sensitive to detect ^{41}Ca in calcium samples enriched (10^{-8}) in ^{41}Ca [15].

The ultimate sensitivity that can be reached is determined either by the background of ^{40}Ca atoms or by the loading rate of the trap. The background of ^{40}Ca atoms can be reduced by improving the isotope selectivity, the loading rate can be increased by improving the throughput of calcium atoms. In this contribution we report on progress in achieving the isotope selectivity necessary for detection of ^{41}Ca at its natural abundance.

2. Experimental progress

At 423 nm, calcium has its resonance transition from the ground $4s^2\ ^1S_0$ state to the $4s4p\ ^1P_1$ state (Figure 1) which is very well suited for laser cooling. The required 423 nm laser light for the cooling is generated by frequency doubling of the output of an 846 nm diode laser system (Toptica Photonics AG). The laser frequency is locked to the cooling transition of calcium by means of polarization spectroscopy [12, 22].

Our ATTA set-up (Al^{41}Catraz) which is schematically depicted in Figure 2 is described in detail elsewhere [10], and we will only mention the main features briefly. An atomic calcium beam is produced by evaporation from an oven (oven 1, Figure 2). In a Zeeman slower [16] atoms are slowed down by a counter-propagating laser beam. The Zeeman slower is designed to decelerate atoms with initial velocities up to 1,000 m/s while the final velocity is 50 m/s. The resulting slow atom beam is deflected over 30° in the direction of a MOT. The deflection

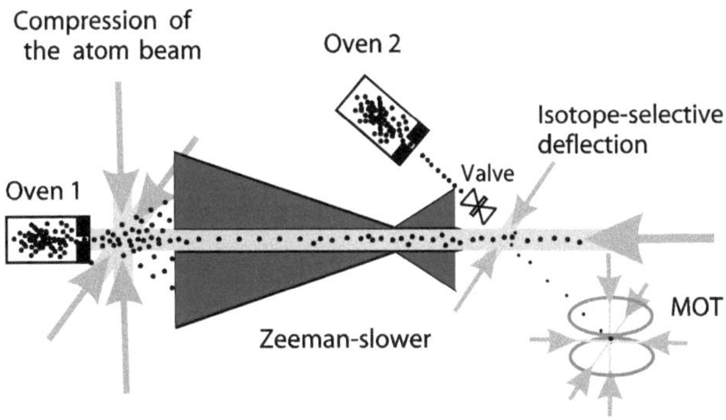

Figure 2. A schematic overview of the experimental setup.

stage as well as the Zeeman slower is highly isotope selective due to the isotope shift. The isotope detection occurs in the MOT. Light scattered by trapped atoms is collected by a large numerical aperture lens system [1] and imaged onto a PMT for single photon counting. To reduce background light, the light from the trap is imaged on a spatial filter before detection by the PMT. To cover the large dynamic range of detecting the various calcium isotopes we employ also a CCD camera as well as a photodiode for the detection of the trap fluorescence.

To show single atom detection in the trap the flux from the oven is reduced down to a loading rate of a few atoms per minute. At that point the fluorescence from the trap displays discrete steps, which allow us to to count the number of atoms in the trap. Figure 3 shows the fluorescence of single ^{40}Ca atoms detected in the trap for a period of 10 s.

The width of the peaks is a measure for the average trapping time of calcium in a MOT. Due to a weak leak (10^{-5}) from the $4s4p\ ^1P_1$ to the $^1D_2$ state, the trapping lifetime is normally limited to $\sim$20 ms [8]. By repumping atoms from the $^1D_2$ to the $5p\ ^1P_1$ state the trapping time can be increased. Atoms in the $5p\ ^1P_1$ state can decay back to the ground state, cf. Figure 1. The laser light required for the repumping is generated by a home built diode laser operating at 672 nm. During the measurement shown in Figure 3 the repump laser was used: we concluded from trap decay time measurements that the average trapping time improved from $\sim$20 ms to $\sim$200 ms. Trapping the atoms longer enables us to use a larger integration time which reduces the statistical error on the signal. For our laser detuning and intensity a trapped atom scatters photons at an estimated rate of $\sim 1.3 \cdot 10^6\ s^{-1}$ of which 35 ± 3 photons per 4 ms are detected. This results in a total photon detection efficiency of $\sim 0.7\%$. The background level produced by stray light of the trapping beams is 50 photons per 4 ms. This is not to be confused with the noise level, which is limited by statistics.

Next to efficiency in single atom detection, throughput is the key issue. Although our oven is designed to produce atomic beams with a small divergence, the divergence is still an important factor in the whole transmission from oven to MOT. The divergence of a specific isotopic componenet of the atomic beam can

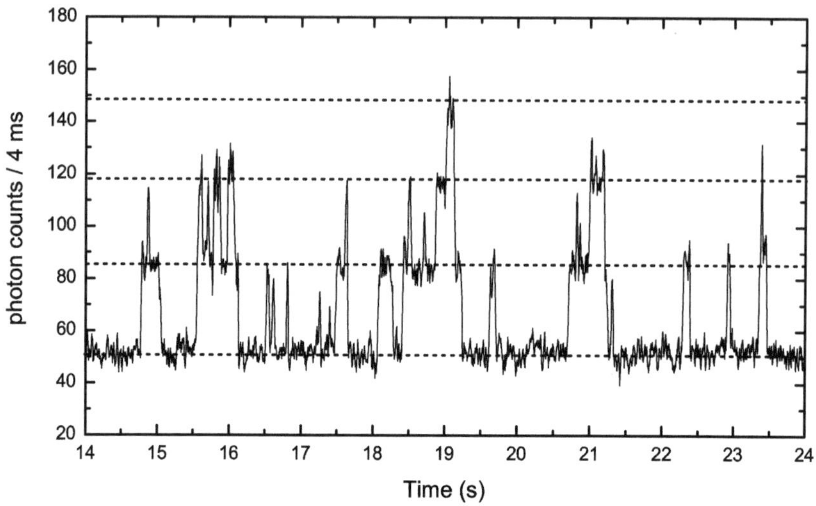

Figure 3. Fluorescence of individual atoms detected in the MOT. The *dotted lines* indicate the fluorescence level for one, two and three atoms in the trap. The integration time of the photon detector is 4 ms. The trapping laser light shining into the MOT chamber is causing the photon background level of ~50 counts/4 ms.

be reduced using optical compression with laser beams perpendicular to the atomic beam [3]. The transverse velocity in the calcium beam is less than one-tenth of the longitudinal velocity, for oven conditions in which the mean free path of the atoms is larger than the length of the exit channel of the oven. The associated Doppler shift, the isotope shift and the natural line width of the transition together determine the effective scattering rate of a specific isotope. The typical isotope shift between two adjacent isotopes of calcium is about 160–200 MHz. The natural line width of the cooling transition is 34 MHz, and the Doppler shift for the calcium atoms is 2.1 MHz/(m/s). We have performed Monte Carlo type of calculations to demonstrate that if the transverse Doppler broadening is limited isotope-selective compression is feasible. Tuning the laser frequency in between two adjacent isotopes, the laser frequency is red-shifted for the heavier isotope and blue-shifted for the lighter isotope therefore the force on both isotopes is opposite in sign [14]. This implies that the lighter isotope will be pushed away from the beam axis while at the same time the transverse velocity component of the heavier isotope can be reduced, resulting in an improved transmission of the heavier isotope.

The outcome of our Monte Carlo simulations is depicted in Figure 4. A laser frequency is choosen which is 60 MHz red-detuned from ^{41}Ca and 90 MHz blue-detuned from the ^{40}Ca resonance. Over a distance of 1 cm, the counterpropagating laser beams interact at right angles with the atomic beam. From the figure it is clear that the divergence of the heavier isotope is strongly reduced while the one of the lighter isotope is increased.

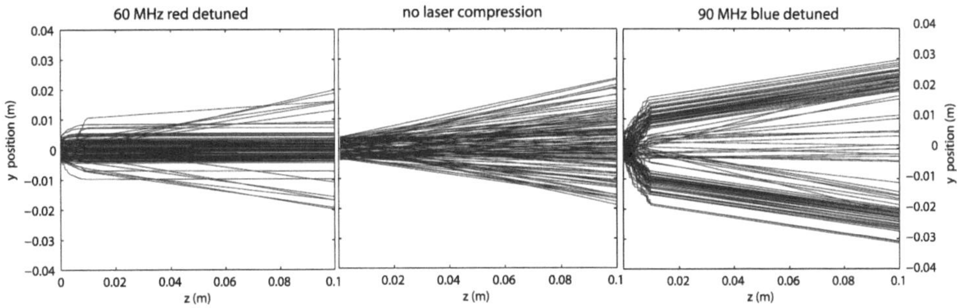

Figure 4. Monte Carlo simulation of the radial position of a beam of calcium atoms *versus* longitudinal distance: *left panel*) ^{41}Ca and *right panel*) ^{40}Ca. The laser beams along the +y and −y direction are choosen to interact with the atomic beam over the first 0.01 m only. For reference the *middle panel* shows the beam divergence without laser compression.

It is of note that the laser power used in the simulations was 20 mW, whilst in the experiment only 5 mW is available, because at present the total output power of the laser is approximately 60 mW. Therefore compression is only achieved partly. On basis of single-atom detection of ^{46}Ca which has a natural abundance of 4×10^{-5} our present sensitivity for ^{41}Ca is estimated to be 1 atom per hour at an abundance level of 10^{-12} Increasing the laser power will increase the efficiency and further improve the performance and the isotope selectivity of the individual components: transversal beam compression, Zeeman slower and deflection stage. Adding a second isotope selective compression stage directly after the Zeeman slower is another potential improvement. With these changes the detection of ^{41}Ca from natural calcium samples becomes within reach.

3. Conclusion

In our Al^{41}Catraz experiment, we aim at the detection of the calcium isotope ^{41}Ca at the natural abundance level of approximately 10^{-14}. Progress has been made in single-atom detection of all stable isotopes of calcium. On basis of single-atom detection of ^{46}Ca our present sensitivity for ^{41}Ca is estimated to be 1 atom per hour at an abundance of 10^{-12}. Monte Carlo simulations show that the throughput can be increased very strongly by beam compression.

Acknowledgements

The authors would like to thank the KVI technical staff for their support and Toptica Photonics for prompt assistance. This project (00PR1887) is part of the research program of the Stichting voor Fundamenteel Onderzoek der Materie (FOM) which is supported by the Nederlandse Organisatie voor Wetenschappelijk Onderzoek (NWO). The work is also supported within the EU network NIPNET (HPRI-CT-2001-50034).

References

1. Alt W., *Optik* **113** (2002), 142.
2. Bailey K., Chen C. Y., Du X., Li Y. M., Lu Z. T., O'Connor T. P. and Young L., *Nucl. Instrum. Methods B* **172** (2000), 224.
3. Balykin V. I., Letokhov V. S., Minogin V. G., Rozhdestvensky Y. V. and Sidorov A. I., *J. Opt. Soc. Am. B* **2** (1985), 1776.
4. Chen C. Y., Li Y. M., Bailey K., O'Connor T. P., Young L. and Lu Z. T., *Science* **286** (1999), 1139
5. Du X., Purtschert R., Bailey K., Lehmann B. E., Lorenzo R., Lu Z. T., Müller P., O'Connor T. P., Sturchio N. C. and Young L., *Geophys. Res. Lett.* **30** (2003), 2068.
6. Freeman S. P. H. T., Beck B., Bierman J. M., Caffee M. W., Heaney R. P., Holloway L., Marcus R., Southon J. R. and Vogel J. S., *Nucl. Instrum. Methods B* **172** (2000), 930.
7. Freeman S., Wendt K., Müller P. and Geppert C., *J. Bone Miner. Res.* **16** (2001), S346.
8. Grünert J. and Hemmerich A., *Appl. Phys. B* **73** (2001), 815
9. Henning W., Bell W. A., Billquist P. J., Glagola B. G., Kutschera W., Liu Z., Lucas H. F., Paul M., Rehm K. E. and Yntema J. L., *Science* **236** (1987), 725.
10. Hoekstra S., Mollema A. K., Morgenstern R., Wilschut H. W. and Hoekstra R., *Phys. Rev. A* **71** (2005), 023409.
11. Kutschera W., Golser R., Priller A. and Strohmaier B. (ed.), *Nucl. Instrum. Methods B* **172** (2000), 1
12. Lancaster G. P. T., Conroy R. S., Clifford M. A., Arlt J. and Dholakia D., *Opt. Commun.* **170** (1999), 79
13. Lu Z. T. and Wendt K. D. A., *Rev. Sci. Instrum.* **74** (2003), 1169.
14. Metcalf H. and Straten P. v. d., *Laser Cooling and Trapping*, Springer, Berlin Heidelberg New York, 1999.
15. Moore I. D., Bailey K., Greene J., Lu Z. T., Müller P., O'Connor T. P., Geppert C., Wendt K. D. A. and Young L., *Phys. Rev. Lett.* **92** (2004), 153002.
16. Phillips W. D. and Metcalf H., *Phys. Rev. Lett.* **48** (1982), 1149
17. Raab E. L., Prentiss M., Cable A., Chu S. and Pritchard D. E., *Phys. Rev. Lett.* **59** (1987), 2631.
18. Raisbeck G. M. and Yiou F., *Nature* **277** (1979), 42
19. Sturchio N. C., Du X., Purtschert R., Lehmann B. E., Sultan M., Patterson L. J., Lu Z. T., Muller P., Bigler T. and Bailey K., *Geophys. Res. Lett.* **31** (2004), 05503.
20. Taylor R. E., *Anal. Chem.* **59** (1987), A317.
21. Wendt K. D. A., Blaum K., Geppert C., Horn R., Passler G., Trautmann N. and Bushaw B. A., *Nucl. Instrum. Methods B* **204** (2003), 325.
22. Wieman C. and Hänsch T. W., *Phys. Rev. Lett.* **36** (1976), 1170.

Hyperfine Interactions (2005) 162:173–179
DOI 10.1007/s10751-005-9223-z

Laser Ionization and Penning Trap Mass Spectrometry – A Fruitful Combination for Isomer Separation and High-precision Mass Measurements

K. BLAUM[1,2,*], D. BECK[2], G. BOLLEN[3], P. DELAHAYE[4], C. GUÉNAUT[5], F. HERFURTH[2], A. KELLERBAUER[4], H.-J. KLUGE[2], U. KÖSTER[4], D. LUNNEY[5], S. SCHWARZ[3], L. SCHWEIKHARD[6] and C. YAZIDJIAN[2,4]

[1] *Institute of Physics, Johannes Gutenberg-University, 55099 Mainz, Germany; e-mail: blaumk@uni-mainz.de*

[2] *GSI Darmstadt, Planckstraße 1, 64291 Darmstadt, Germany*

[3] *NSCL, Michigan State University, East Lansing, MI 48824-1321, USA*

[4] *Department of Physics, CERN, 1211 Geneva 23, Switzerland*

[5] *CSNSM-IN2P3-CNRS, 91405 Orsay-Campus, France*

[6] *Institute of Physics, Ernst-Moritz-Arndt University, 17487 Greifswald, Germany*

Abstract. We have demonstrated for the first time that element-selective laser ionization in combination with ultra-high resolution mass spectrometry can be used to prepare isomerically pure ion ensembles. Together with β–γ coincidence studies this method allowed a determination of the low-energy structure and the unambiguous identification of triple β-decaying isomerism in ^{70}Cu. By selective resonant ionization and measurement of the masses of these three states using ISOLTRAP at ISOLDE/CERN with a relative uncertainty of $\delta m/m \approx 5 \cdot 10^{-8}$ a clear state-to-mass assignment was possible which resolved the assignment puzzle in ^{70}Cu.

Key Words: laser ionization, mass spectrometry, nuclear isomers, nuclear masses, penning trap.

1. Introduction

Almost one third of the nuclides in the nuclear chart have long-lived excited states with – in many cases – unknown excitation energies [1]. The resolution of isomeric states is therefore an important issue in direct mass measurements on short-lived radionuclides. To achieve this and to solve identification problems, often a resolving power of well above 10^6 is needed. By combination of element-selective laser ionization and high-resolution mass spectrometry we demonstrated not only resolution but also isolation of ground and excited states and prepared isomerically pure ensembles of ^{68}Cu and ^{70}Cu. To this end, the resonant ionization laser ion source (RILIS) [2] and the triple-trap mass-spectrometer ISOLTRAP [3–5] at ISOLDE/CERN [6] were used. Together with β–γ coincidence studies this method allowed to determine the low-energy structure

---

* Author for correspondence.

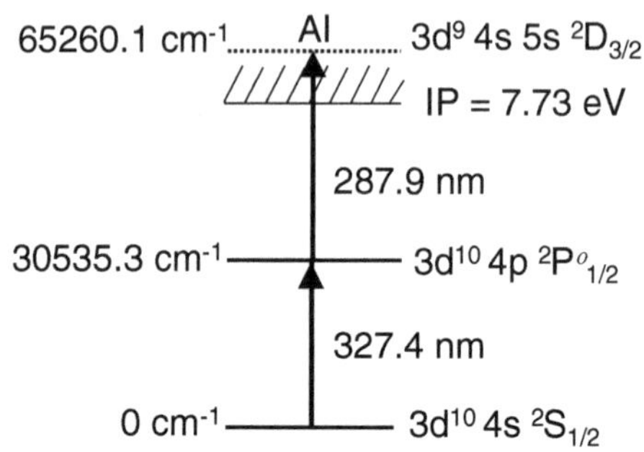

Figure 1. Excitation scheme for double-resonance ionization of copper.

of ^{70}Cu and to unambiguously identify three β-decaying isomers. By two-step resonant laser ionization and high-resolution Penning trap mass measurements it was possible to perform high-precision mass determinations on each state. Thus, a clear identification was possible, which allowed to solve the assignment problem in ^{70}Cu [7].

2. Experimental procedure

2.1. PRODUCTION AND IONIZATION OF SHORT-LIVED COPPER ISOTOPES AT ISOLDE

Copper atoms were produced at the on-line isotope separator ISOLDE by bombarding a 54-g/cm^2 UC_x/graphite target with 1.4-GeV proton pulses from the CERN Proton Synchrotron Booster containing about $3 \cdot 10^{13}$ protons each. The copper isotopes diffused out of the target, which is heated to 2000°C, and were selectively laser ionized with the RILIS [8]. The ions were accelerated to 60 keV and mass-separated in the general-purpose mass separator (GPS). The GPS was operated at a mass resolving power of $m/\delta m \approx 1000$. The Cu^+ beam with a typical intensity of the order of 10^7 ions/s in the case of ^{68}Cu and 10^5 ions/s for ^{70}Cu was guided to the tandem Penning-trap mass spectrometer setup.

For the laser ionization an efficient two-step resonant excitation and ionization scheme was used. The atomic excitation path for the copper isotopes is shown in Figure 1. It was a resonant excitation followed by ionization through a strong autoionizing (AI) state. To this end, two copper vapor lasers with a repetition rate of 11 kHz pumped the frequency-doubled dye lasers. The bandwidth of the laser for the first transition at $\lambda = 327.4$ nm was 1.2 GHz, which made it possible to resolve the hyperfine structures of the ^{68,70}Cu isomers [9, 10]. For the ionization step a broadband ($\approx$ 24 GHz) dye laser was used.

2.2. HIGH-PRECISION MASS MEASUREMENTS WITH ISOLTRAP

The mass measurements on the copper isotopes were performed with the Penning-trap mass-spectrometer ISOLTRAP [3–5]. The apparatus (Figure 2)

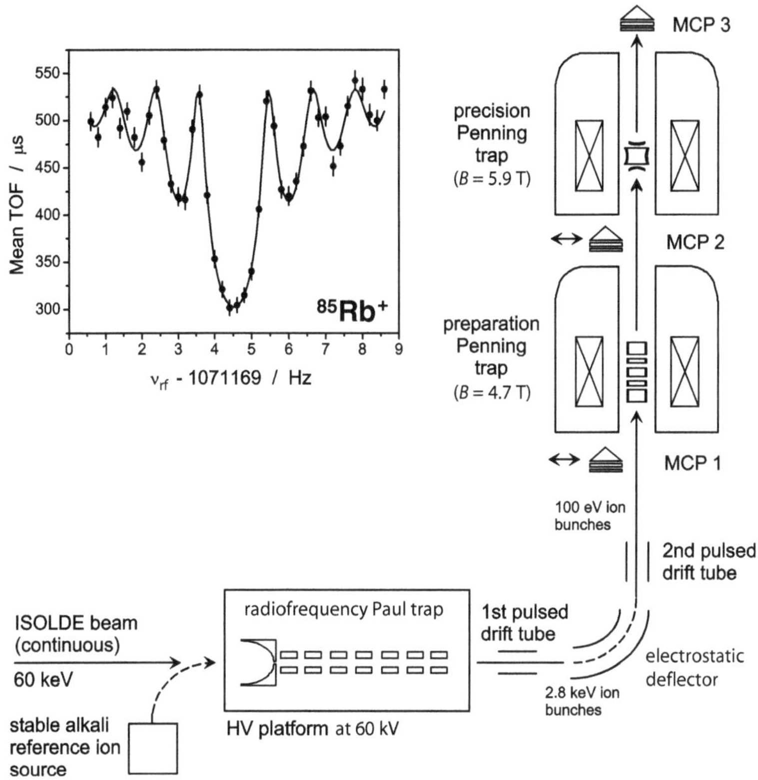

Figure 2. Schematic drawing of the mass spectrometer ISOLTRAP including the radiofrequency Paul trap and the preparation and precision Penning traps. Micro-channel-plate (MCP) detectors are used to monitor the ion transfer (MCP1 and MCP2) as well as to record the time of flight (MCP3) for the determination of the cyclotron frequency. The *inset* shows a resonance of $^{85}Rb^+$ for the magnetic-field calibration. The *line* is a fit of the theoretical curve to the data points [13].

consists of a gas-filled linear radiofrequency Paul trap for beam preparation [11], a cylindrical preparation Penning trap for cooling and isobaric cleaning of the ion bunch, and a hyperbolical Penning trap for the precision mass measurements. A mass measurement is carried out via the determination of the cyclotron frequency $\nu_c = qB/(2\pi m)$ of the ion stored in the precision Penning trap, where q/m is the charge-to-mass ratio of the ion and B is the magnetic-field magnitude. The two superconducting magnets have field strengths of 4.7 T (preparation trap) and 5.9 T (precision trap), respectively. ν_c is probed by excitation of the ions' motion with a time-varying quadrupolar electric field of frequency ν_{rf}. At $\nu_{rf} = \nu_c$, a full conversion from initially pure magnetron motion to cyclotron motion is obtained [12, 13] and the orbital magnetic moment μ and hence the radial kinetic energy $E_r = \mu B$ are maximal. Since the axial acceleration of the ions in the fringe field of the magnet is proportional to μ, the ions' time of flight (TOF) to the micro-channel-plate detector MCP3 is a measure of the radial energy. The TOF cyclotron resonance (see inset of Figure 2) is determined by repetition of this sequence and measurement of the TOF as a function of the frequency of the applied field. The minimum of the resonance is at $\nu_{rf} = \nu_c$. The magnetic field strength B is determined via the cyclotron frequency of $^{85}Rb^+$ with well-known mass.

The Fourier limit of the resolving power for a given excitation time T_{rf} is: $R = m/\Delta m = \nu_c/\Delta\nu_c(\text{FWHM}) \approx 1.25 \cdot \nu_c \cdot T_{rf}$ [14]. Thus, $R \approx 10^6$ is reached with ISOLTRAP for singly charged $A \approx 100$ ions in the precision Penning trap and an excitation time of $T_{rf} \approx 1$ s (for sufficiently long half-lives). Even higher resolving powers can be obtained by further increasing the excitation time. Cross-reference measurements on carbon clusters have shown that the overall accuracy of ISOLTRAP for the frequency ratio of two ions is $8 \cdot 10^{-9}$ [15]. In total, the masses of close to 300 radionuclides have been measured by use of ISOLTRAP with a relative uncertainty of typically $\delta m/m = 10^{-8} - 10^{-7}$ [16].

2.3. ISOLATION OF ISOMERICALLY PURE ENSEMBLES

Due to the broad bandwidth and the Doppler broadening and thus the overlapping hyperfine structure the selectivity of the RILIS was not sufficient to separate the ^{70}Cu isomers effectively. A cyclotron excitation time of $T_{rf} = 0.9$ s was used in the precision Penning trap. This results in a resolving power of more than 10^6, sufficient to clearly resolve the ^{68,70}Cu isomers. In order to avoid systematic errors it is not only important to resolve ions with different masses but also to selectively remove unwanted species, as for instance remaining isobaric contaminations or isomers. This can be achieved by mass-selective radial ejection (via a dipolar excitation at the reduced cyclotron frequency) of the unwanted ions, i.e., ions belonging to the unwanted nuclear state. In the experiments presented, radiofrequency ejection durations of 3 s were used to prepare isomerically pure ensembles. In addition, the number of ions per cycle in the precision trap was kept low (1–5) in order to minimize frequency shifts caused by ion–ion interactions.

3. Results and discussion

By scanning the laser frequency of the first transition in the resonant-ionization scheme [9] and simultaneously performing β–γ coincidence measurements [17] the radionuclides were detected and identified. Investigation of the intensities of the individual γ rays in the β decay of ^{70}Cu revealed distinct groups of γ rays belonging to three different hyperfine-structure patterns as shown in the top part of Figure 3. This is a clear evidence for the existence of three β-decaying isomers in ^{70}Cu. Spin values were deduced from the magnetic moments with spin (6^-) for the lowest, (3^-) for the intermediate and 1^+ for the highest-lying isomer [10].[1]

[1] Tentative spin values are given with parentheses, known spin values without.

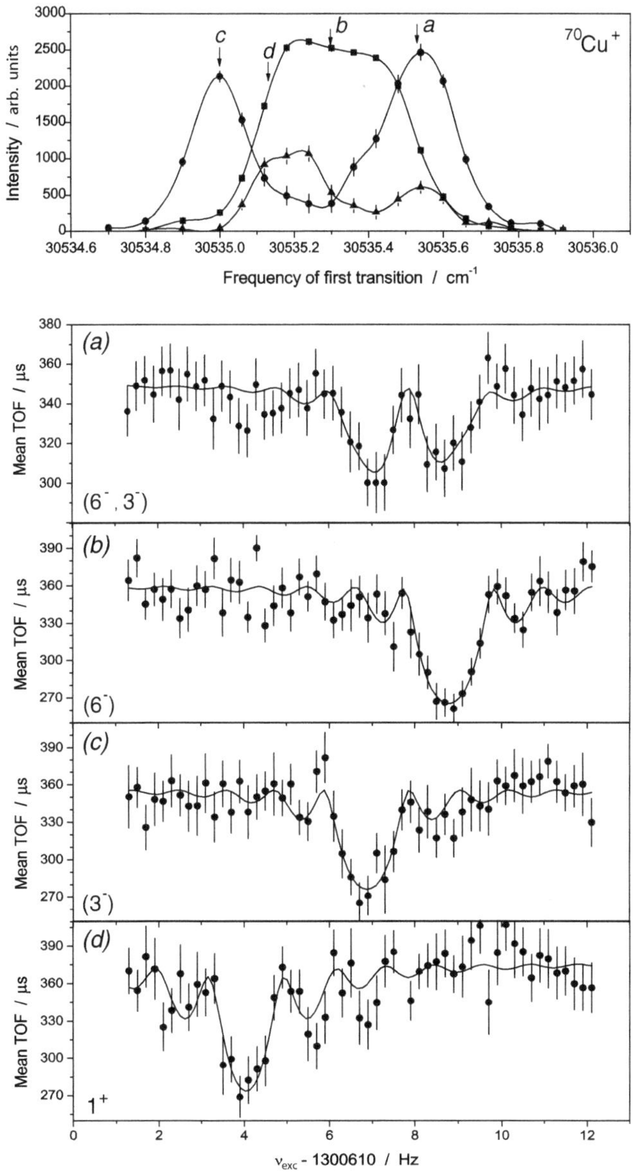

Figure 3. Top: Measured (β–γ) radioactivity of the ground (*squares*), first (*bullets*) and second (*triangles*) excited state of $^{70}Cu^+$ *versus* laser frequency of the first transition. *Bottom*: Time-of-flight cyclotron resonance curves as a function of excitation frequency for the laser settings marked with *arrows* (a, b, c, d) in the top (a) shows the isomers resolved but not separated while (b)–(d) shows the *separation* of the isomeric states. The assignment of the different spin states to the isomeric states results from the ISOLTRAP measurement (see text). The tentative spin values were taken from [10]. The *solid lines* are fits of the theoretically expected line shape [13] for two (a) or one (b)–(d) resonance curves to the data points.

To perform high-precision mass measurements on the three different isomeric states the laser frequency was tuned to the positions (a, b, c, d) as indicated by arrows in the upper part of Figure 3. The obtained TOF cyclotron resonances are shown in the lower part of Figure 3. For the laser frequencies set to position a and d the selectivity of the RILIS was not sufficient to prepare isomerically pure samples. Since the resolving power of ISOLTRAP is high enough to clearly separate the different resonances (see (a) in the lower part of Figure 3), we applied the abovementioned mass-selective cleaning procedure to one or two of the isomers to obtain an isomerically clean cyclotron resonance of the isomer of interest (see resonance shown in part (d)). For the positions b and c the selectivity of the laser ionization was sufficient to obtain almost pure ensembles of the (6^-) and (3^-) isomer (see part (c) and (d) in Figure 3) and no additional cleaning in the precision trap was applied.

Since the cyclotron frequency is inversely proportional to the mass of the ion, our results unambiguously determined the order of the levels and assigned spins to states, i.e., the (6^-) is the ground state, the (3^-) the first, and the 1^+ the second excited state, as proposed in [10]. The mass differences are in excellent agreement with the excitation energies obtained in the decay studies and are given in [7]. The same procedure was applied to ^{68}Cu to prepare pure ensembles of the ground and excited state and to perform background-free mass measurements [18].

Acknowledgements

The authors thank the ISOLDE staff for their continuous support. This work was supported by the German Ministry for Education and Research (BMBF) under contract number 06GF151, by the European Commission under contracts IHRP HPRI–CT–1999–00018, HPRI–CT–2001–50033 (TARGISOL), HPRI–CT–2001–50034 (NIPNET), and HPMT-CT-2000-00197 (Marie Curie Fellowship), and by the Helmholtz Association of National Research Centres (HGF) under contract number VH-NG-037.

References

1. Audi G., Wapstra A. H. and Thibault C., *Nucl. Phys. A* **729** (2003), 337.
2. Fedoseyev V. N. *et al., Hyperfine Interact.* **127** (2000), 409.
3. Bollen G. *et al., Nucl. Instr. Methods A* **368** (1996), 675.
4. Herfurth F. *et al., J. Phys. B* **36** (2003), 931.
5. Kellerbauer A., *Int. J. Mass Spectrom.* **229** (2003), 107.
6. Kugler E., *Hyp. Int.* **129** (2000), 23.
7. Van Roosbroeck J. *et al., Phys. Rev. Lett.* **92** (2004), 112501.
8. Köster U. *et al., Spectrochim. Acta B* **58** (2003), 1047.
9. Köster U. *et al., Nucl. Instr. Methods B* **160** (2000), 528.
10. Weissman L. *et al., Phys. Rev. C* **65** (2002), 024315.

11. Herfurth F. *et al., Nucl. Instr. Methods A* **469** (2001), 254.
12. Gräff G., Kalinowsky H. and Traut J., *Z. Phys. A* **297** (1980), 35.
13. König M. *et al., Int. J. Mass Spectrom. Ion Processes* **142** (1995), 95.
14. Bollen G., *Nucl. Phys. A* **693** (2001), 3.
15. Kellerbauer A. *et al., Eur. Phys. J. D* **22** (2003), 53.
16. Blaum K. *et al., Nucl. Phys. A* **746** (2004), 305c.
17. Van Roosbroeck J. *et al., Phys. Rev. C* **69** (2004), 034313.
18. Blaum K. *et al., Europhys. Lett.* **67** (2004), 586.

Hyperfine Interactions (2005) 162:181–188
DOI: 10.1007/s10751-005-9217-x

Laser Cooling and Spectroscopy of Relativistic C^{3+} Beams at the ESR★

U. SCHRAMM[1,*], M. BUSSMANN[1], D. HABS[1], M. STECK[2], T. KÜHL[2], K. BECKERT[2], P. BELLER[2], B. FRANZKE[2], F. NOLDEN[2], G. SAATHOFF[3], S. REINHARDT[3], and S. KARPUK[4]
[1]*Department für Physik, LMU München, D-85748 Garching, Germany; e-mail: uschramm@garching.physik.uni-muenchen.de*
[2]*Gesellschaft für Schwerionenforschung (GSI), D- Darmstadt, Germany*
[3]*Max-Planck-Institut für Kernphysik, D-Heidelberg, Germany*
[4]*Institut für Physik, Universität Mainz, D-Mainz, Germany*

Abstract. We report on the first laser cooling of a bunched beam of multiply charged C^{3+} ions performed at the ESR (GSI) at a beam energy of $E = 1.47$ GeV. Moderate bunching provided a force counteracting the decelerating laser force of one counterpropagating laser beam. This versatile type of laser cooling lead to longitudinally space-charge dominated beams with an unprecedented momentum spread of $\Delta p/p \approx 10^{-7}$. Concerning the beam energy and charge state of the ion, the experiment depicts an important intermediate step from the established field of laser cooling of ion beams at low energies toward the unique laser cooling scheme proposed for relativistic beams of highly charged heavy ions at SIS 300 (FAIR).

Key Words: crystalline ion beams, laser cooling, Li-like heavy ions, storage rings.

1. Introduction

The cooling of stored heavy ion beams to high phase space densities is of general importance when low momentum spread is required for precision spectroscopy or high luminosity for collision experiments. Ultimately, cooling the thermal energy far below the mutual Coulomb energy of the ions leads to a phase transition – a crystallization of the ion beam [1] – into a regime where almost no further heating occurs and maximum brilliance is reached [2].

In the low energy regime this phase transition could recently be observed with laser-cooled $^{24}Mg^{+}$ ion beams [3–5]. However, as laser cooling relies on the repeated resonant scattering of photons, only a very limited number of ions can be accessed by this powerful cooling technique at existing storage rings, benchmarking and present activities being summarized in [2].

★ Funded by the German BMBF under contract number 06ML183.
* Author for correspondence.

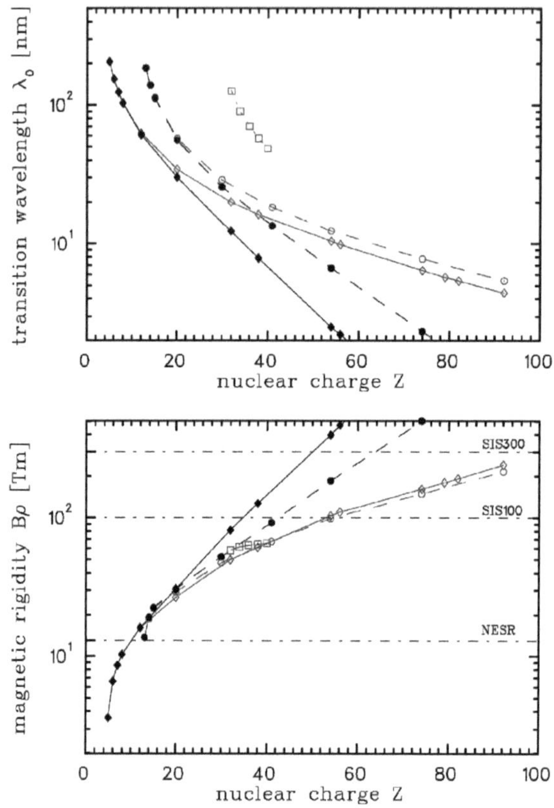

Figure 1. Left graph: Wavelength of the $nS_{1/2} - nP_{1/2}$ (*black lines*) and $nS_{1/2} - nP_{3/2}$ (*grey lines*) ground state transitions in the ion rest frame as a function of the nuclear charge [11]. The *solid lines* (*rhombs*, $n = 2$) refer to Li-like ions, the *dashed lines* (*circles*, $n = 3$) to Na-like and the *dotted lines* (*squares*, $n = 4$) to Cu-like ions. *Right graph*: Corresponding magnetic rigidity required for the storage of the ions at an energy Doppler shifting the transition wavelength to a laser wavelength of $\lambda_{uv} = 257$ nm. The *horizontal lines* denote the maximum rigidity of the FAIR storage rings.

At heavy ion storage rings like ESR (GSI), electron cooling has developed into a versatile tool for the cooling of highly charged ions independent from their internal atomic properties [6]. As the cooling force increases with the square of the ion charge, beam ordering effects of highly charged ions were observable with electron cooling at very low beam currents [7, 8], where competing heating mechanisms are reduced.

In the relativistic regime, that will be accessible at the future FAIR facility, the situation might be reversed. Electron cooling as the established tool in heavy ion storage rings cannot be readily applied any more. Yet, laser cooling, up to now only applied to beams of singly charged ions, seems promising. Due to the fast transition rates in highly charged ions and the relativistically increased momentum transfer in the photon scattering, the cooling force is predicted to principally increase with the third power of the ion energy [2, 9]. Moreover, ground-state transitions of all Li- and most Na-like heavy ions can be reached at FAIR when counterpropagating laser and ion beams are used as demonstrated in Figure 1 [10].

As an intermediate step between the laser cooling of highly relativistic heavy ion beams, anticipated at FAIR [10], and the known field of low energy beams [2], a test experiment was performed with a C^{3+} ion beam at a velocity of $\beta = 0.47$ at the ESR (GSI), of which first results will be presented.

2. Laser cooling of Li-like carbon ions

At the heavy ion storage ring ESR (GSI) laser cooling of Li-like C^{3+} ions becomes possible at a beam energy of 122 MeV/u ($\beta = 0.47$, $\gamma = 1.13$). At this energy the closed $2S_{1/2} - 2P_{3/2}$ transition ($\lambda_0 = 154.82$ nm, $\tau = 3.8$ ns [11]) is Doppler-shifted into resonance with the frequency doubled line of the Ar-ion laser ($\lambda_{laser}/2 = 257.34$ nm) when counterpropagating laser and ion beams are used. The decelerating laser force is counteracted by the restoring force of a the bucket when the beam is bunched. This technique provides the momentum dependent friction force required for cooling [2] without the need of a copropagating laser beam.

In order not to challenge the laser force in this first experiment moderate bunching voltages of only few volts were applied at the 10th as well as at the 20th harmonic of the revolution frequency $f_{rev} = 1.295$ MHz. The bucket depth was determined by the measurement of the synchrotron frequency $f_{sync} \sim 100$ Hz ($h = 10$) and $f_{sync} \sim 170$ Hz ($h = 20$) and corresponds to a momentum acceptance of the order of $\Delta p/p \sim 2 \times 10^{-5}$ ($h = 10$). For an electron pre-cooled beam this momentum spread is reached for a total ion number of few 10^7 (few 10 μA) limited by heating due to intra-beam scattering [2].

Bunched beam laser cooling relies on the damping of the single particle synchrotron motion by the strong but narrow-band resonant laser force. For a typical single-mode laser the width of the force is determined by the line-width of the transition of the ion used for the cooling and not by the width of the laser. It corresponds to $\Delta p/p \approx 5 \times 10^{-8}$. The width of the force is thus extremely mismatched to the initial momentum distribution, yet it principally allows cooling to similarly low momentum spreads. Two schemes are reported so far to resonantly interact with the full ion momentum distribution (see [2, 12]). For the first, invented at ASTRID (Aarhus) [13], the bunching frequency is continuously tuned from a starting frequency where the laser is resonant with ions just at the edge of the bucket to a final frequency close to the center of the bucket. This scheme is illustrated in Figure 2 and described in the caption. The final tuning has to be done carefully as crossing the center of the bucket leads to an equally strong driving of the synchrotron motion, visible in the lower half of Figure 2. The second scheme, invented at TSR [14, 15], relies on the repeated interaction

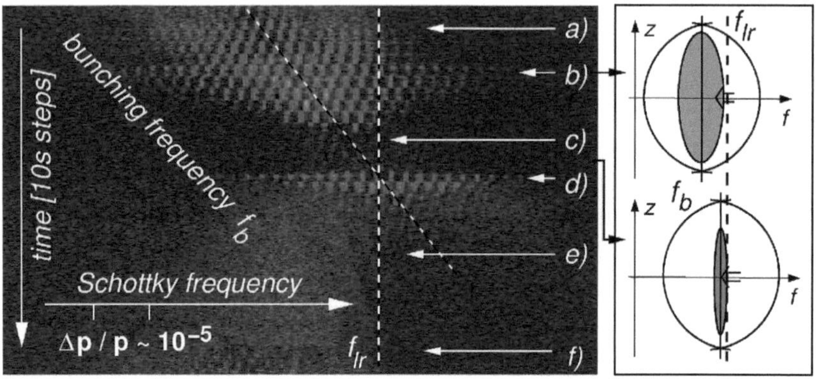

Figure 2. Schottky noise analysis (at the 47th harmonic) of a laser cooled bunched beam of C^{3+} ions with rising bunching frequency f_b (*diagonal dashed line*). The decelerating laser beam is operated at fixed frequency and resonant with ions of the same momentum class f_{lr} (*vertical dashed line*). The bunching frequency $f_b = hf_{rev} = 10 \times 1.295$ MHz is increased in steps of 10 Hz every 10 s. Starting at a low bunching frequency where ions close to the separatrix of the bucket come into resonance with the cooling laser beam, the ion beam is cooled into the bucket (a). Within the harmonic range, the number of synchrotron side bands increases (b and *upper sketch*) interpreted as a laser cooled bunched beam where individual ions perform synchrotron oscillations with $f_{sync} \sim 100$ *Hz*. From the synchrotron frequency (*the side band spacing*) a momentum acceptance of about $\Delta p/p \sim 2 \times 10^{-5}$ can be deduced that corresponds to the observed momentum spread (roughly *the number of side bands*). Closer to resonance ($f_b \rightarrow f_{lr}$) the side bands suddenly vanish (c) when the beam enters the space-charge dominated regime. When the laser beam comes into resonance with ions at the bucket center (d) the synchrotron motion is driven instead of damped. Ions are decelerated (e) out of the bucket until the cycle restarts (f).

of the ions circulating in the bucket with the laser tuned to optimum cooling close to the center of the bucket. Both schemes have been systematically studied at the ESR and detailed results will be published elsewhere.

In Figure 3a a complementary view of the first scheme is given. For decreasing detuning Δf_b and thus for increasing cooling strength for ions around the bucket center, the spatial width of the bunches, measured using capacitive pick-up devices, is shown. Following a continuous reduction of the bunch length, the length remains constant for $\Delta f_b < 100$ Hz. At the same detuning of about 100 Hz (corresponding to $\Delta p/p = 6 \times 10^{-6}$ for $h = 20$) the side bands in a Schottky noise spectrum similar to the one presented in Figure 2 abruptly vanish, interpreted above as the transition to a space-charge dominated regime. In this regime the length of the bucket does not depend on the momentum spread any more, being exactly the signature observed in the bunch length measurement depicted in Figure 3a. No ions are lost at this point although the Schottky-signal appears to be considerably weaker, as the integrated pick-up signal (not shown) remains constant. Furthermore, the measured bunch length can be reproduced under the assumption of space charge dominated bunches within 10% [16].

In the space-charge dominated regime, the momentum spread cannot be derived from the spatial distribution. However, laser cooling itself provides a

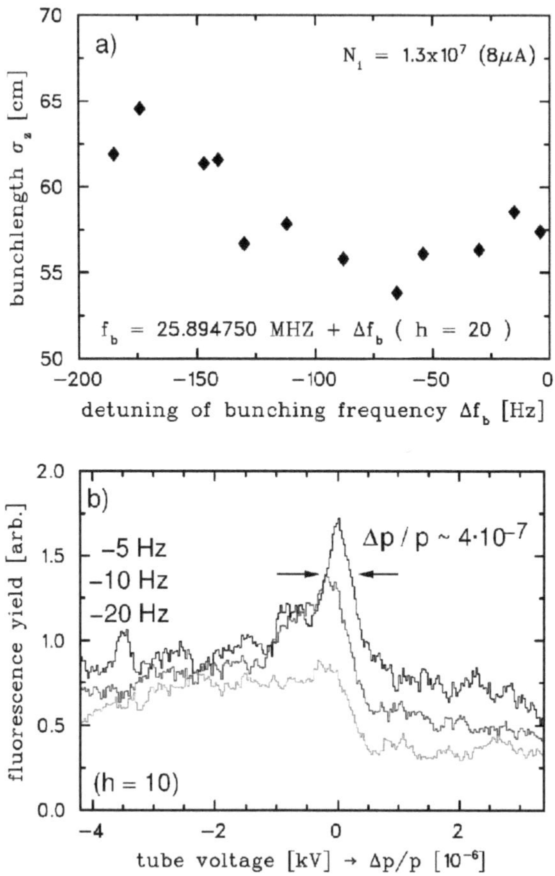

Figure 3. (a) Bunch length of a laser-cooled beam as a function of the detuning Δf_b of the bucket center with respect to the laser resonance. At the detuning for which the side band structure vanishes in the corresponding Schottky spectrum ($\Delta f_b \sim 100$ Hz, here $h = 20$) the width of the bunch remains constant, indicating the space-charge dominated regime. (b) Momentum distribution of the laser cooled bunched beam ($h = 10$) close to minimum detuning. The momentum distribution is measured *via* the Doppler effect making use of a local change of the ion momentum inside a drift tube where the fluorescence is recorded, when the tube voltage is ramped.

unique diagnostic. Tuning the laser frequency over the Doppler-broadened line, the momentum distribution can be directly measured. In practice, the ion momentum is locally tuned by ramping a drift-tube. The result is depicted in Figure 3b. For a frequency detuning of only 5 Hz (corresponding to $\Delta p/p = 6 \times 10^{-7}$ at $h = 10$) a momentum spread of $\Delta p/p = 4 \times 10^{-7}$ is measured for a cold fraction of the beam. It is likely that part of the beam is about one order of magnitude hotter due to intra-beam scattering and subsequent re-circulating in the bucket, yet this beam represents the coldest measured in the ESR. For an ion number of about 1.5×10^6 it corresponds to a plasma parameter of $\Gamma \lesssim 1$.

Summarizing the complete experiment on laser cooling of Li-like carbon ions at 1.47 GeV, the full momentum acceptance of the test-bucket could be laser cooled without prior electron cooling. Yet, the transverse cooling of the injected beam only worked reasonably well after few seconds of electron pre-cooling. Otherwise, the motion in the different degrees of freedom seemed to have decoupled and the beam remained transversally hot. When cooled, the transverse profile was comparable to that of an electron cooled beam. Though, in principle, the laser cooling force exceeds the electron cooling force, electron cooling overrode laser cooling. This observation can be attributed to the extremely narrow bandwidth of the force and (in this special case) also to the insufficient laser intensity, as the transition could only be saturated by $\sim 10\%$.

For efficient laser cooling of the whole ion bunch, the width of the laser force has to be adapted at least to the momentum range that is given by intra-beam scattering, a problem well known from lower energy experiments [17, 18]. This is planned to be realized either by a second scanning laser system or by the use of a pulsed laser systems with pulse length of the order of few 100 ps to 1 ns in the near future.

3. Laser spectroscopy of Li-like carbon ions

As laser cooling of Li-like heavy ions relies on the resonant excitation of the $2S_{1/2} - 2P_{1/2}$ or the $2S_{1/2} - 2P_{3/2}$ transitions, the experiment incorporates the spectroscopy of these transitions. The Ar-ion laser can – in principle – be locked to a calibrated iodine line ($\lambda_{a3} = 514.6734664$ nm) with a relative precision of better than 10^{-9} [19]. However, due to the huge Doppler-effect $\omega_0 = \gamma(1+\beta)\omega_{uv} \approx 1.66\omega_{uv}$ that enables the excitation of transitions in the deep UV or even X-ray range with near UV laser beams, the knowledge of the ion energy determines the accuracy of the spectroscopy in the rest-frame $\Delta\omega_0/\omega_0 = (1-\beta^2)^{-1}\Delta\beta = \beta^{-1}\Delta\gamma/\gamma$ and thus a reliable method for the determination of the ion energy has to be found.

From the Schottky-noise spectra, like the one presented in Figure 2, the revolution frequency of a sufficiently cold ion beam can be deduced with comparative precision. However, as the length of the closed orbit depends on the individual setting of the storage ring lattice (usually known to 3×10^{-4} for the design orbit), this measurement does not translate into a beam energy of equivalent precision. At the ESR, where the beam can be electron-cooled, the determination of the acceleration voltage of the electron beam ($U_e \sim 67$ kV), calibrated with a relative precision of 10^{-4} and showing much less jitter, currently represents the method of choice. The spectroscopy measurement was performed in a way that, first, the position of the laser resonance was marked in the Schottky-spectrum for a coasting uncooled beam with a relative accuracy of $\sim 10^{-7}$. The laser wavelength was stabilized to the low-wavelength side of the

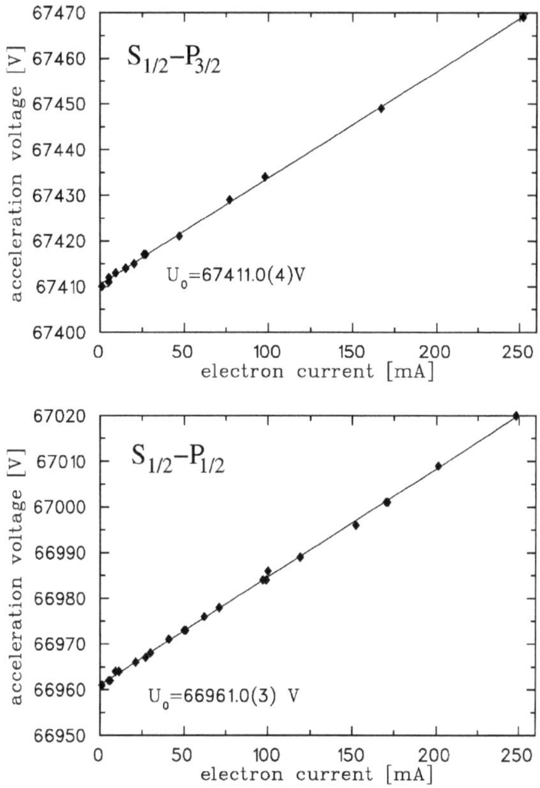

Figure 4. Acceleration voltage of the electron cooler required to match the laser resonance as a function of the electron current.

Doppler-broadened iodine-line at $\lambda_{laser} = 514.6729(3)$ nm, the relative jitter was reduced to $\sim 5 \times 10^{-9}$. Then, electron cooling was activated and tuned to match the original position of the laser-resonance in the Schottky-spectrum to $\sim 10^{-5}$ (digitizing resolution). As the electron energy in the center of the beam, where the ion beam is positioned, is reduced by about 0.1% due to the space charge of the electron beam, the best way of eliminating this correction is to perform the above procedure for different electron currents and to extrapolate to zero current (Figure 4).

	$\lambda(2S_{1/2} - 2P_{1/2})$ [nm]	$\lambda(2S_{1/2} - 2P_{3/2})$ [nm]
Kim *et al.* 1991 [20]	155.060	154.804
Johnson *et al.* 1996 [11]	155.078	154.819
Tupitsyn and Shabaev 2003, private communication [2004]	155.0739(26)	154.8173(53)
This work	155.0705(39)(3)	154.8127(39)(2)
Edlen *et al.* 1983 [21]	155.077	154.820

From these electron energies eU_0, matching the ion energies where the ions are in resonance with the laser light, the following transition wavelengths can be deduced, where the first error denotes the calibration accuracy of the power supply and the second the statistical error.

The present experimental results reach the same accuracy that is set by the actual theoretical work (Tupitsyn and Shabaev 2003, private communication [2004]) and are consistent within one standard deviation. Yet, for both lines the experiment yields slightly lower values and it seems likely that this is due to the absolute calibration of the acceleration voltage and thus a systematic effect.

Summarizing, already for this test experiment the same accuracy was reached that can be set by theory. With an improvement of the voltage calibration or measurements at different ion energies and laser wavelength on the same transition, theory could be readily challenged. At the ESR the method could be extended up to O^{5+} ions, while at FAIR, every element can be reached.

References

1. Schiffer J. P. and Kienle P., *Z. Phys.* A **321** (1985), 181.
2. Schramm U. and Habs D., *Progress in Particle and Nuclear Physics* **53** (2004), 583–677.
3. Schätz T., Schramm U. and Habs D., *Nature* (London) **412** (2001), 717.
4. Schramm U., Schätz T. and Habs D., *Phys. Rev.* **E 66** (2002), 036501.
5. Schramm U. *et al.*, *Journal of Physics* **36** (2003), 561.
6. Steck M. *J. Opt. Soc. Am.* B **20** (2003), 1016.
7. Steck M. *et al.*, *J. Phys.* B **36**, 991 (2003) and *Phys. Rev. Lett.* **77** (1996), 3803.
8. Danared H. *et al.*, *J. Phys.* B **36** (2003), 1003.
9. Schramm U., Bussmann M. and Habs D., *Nucl. Instrum. Methods*, **A 532** (2004), 348.
10. Schramm U. *et al.*, LoI #18 FAIR APPA–PAC (2004).
11. Johnson W. R. *et al.*, *At. Data Nucl. Data Tables* **64** (1996), 279.
12. Schramm U., Schätz T. and Habs D., *Phys. Rev. Lett.* **87** (2001), 184801.
13. Hangst J. S. *et al.*, *Phys. Rev. Lett.* **74** (1995), 4432.
14. Miesner H.-J. *et al.*, *Nucl. Instr. Meth.*, Letter to the Editor, **A 383** (1996), 634.
15. Eisenbarth U. *et al.*, *Nucl. Instrum. Methods*, **A 441** (2000), 209.
16. Ellison T. J. P. *et al.*, *Phys. Rev. Lett.* **70** (1993), 790.
17. Wanner B. *et al.*, *Phys. Rev.* **A 58** (1998), 2242.
18. Atutov S. N. *et al.*, *Phys. Rev. Lett.* **80** (1998), 2129.
19. Saathoff G. *et al.*, *Phys. Rev. Lett.* **91** (2003), 190403.
20. Kim Y.-K. *et al.*, *Phys. Rev.* **A 44** (1991), 148.
21. Tupitsyn I. I. and Shabaev V. M., priv. com. (2004).
22. Edlen B., *Phys. Scripta* **28** (1983), 51.

Hyperfine Interactions (2005) 162:189–194
DOI 10.1007/s10751-005-9220-2

A Solid Xenon Catcher for Rare Isotope Laser Spectroscopy

A. E. EZWAM and J. BILLOWES*
Schuster Laboratory, University of Manchester, Manchester M13 9PL, UK; e-mail: j.billowes@manchester.ac.uk

Abstract. Solid xenon layers are proposed as an alternative to graphite catchers for collecting samples at on-line separators. The formation of solid xenon layers is described. Sample atoms of bismuth have been held for up to a day in such layers before being released as cold free atoms for laser spectroscopy measurement. An application to studying daughter nuclei produced by α-decay is considered. It involves a two-stage process where daughter recoils are first accumulated in the xenon layer and subsequently released into flowing helium for laser-ionization and counting.

1. Introduction

A new technique has been tested which should improve the sensitivity of laser spectroscopic measurements on rare or radioactive isotopes. The aim of this work is to replace conventional graphite catchers, which are used to collect and accumulate isotope samples, with a layer of solid xenon. Advantages of the technique include a simpler thermal release of the sample, a better spatial and temporal overlap of the sample with pulsed laser beams used to study the atoms, and a reduced Doppler broadening of the laser-resonance signal. In the present work the Doppler broadening was reduced by a factor of two compared with room temperature broadening.

Graphite is conventionally used as an ion catcher in techniques developed for ultra-sensitive detection of trace isotopes. See, for example, the work by Krönert *et al.* [1] and the COMPLIS programme carried out at ISOLDE [2]. In the two-stage technique the sample is first implanted in the surface of a graphite catcher using a low-energy ion beam (20–60 keV). The sample atoms may be produced in the ion source by nuclear reactions, or could be introduced into a plasma ion source as material from environmental samples. The extracted ion beam is mass-analysed before the implantation stage. The sample accumulates in the graphite throughout the implantation.

---

* Author for correspondence.

In the second stage the sample is released by laser ablation and the atoms are typically observed by laser resonance ionization spectroscopy (RIS) methods. The brutal nature of the release makes this part of the technique inefficient. The ablation process locally heats the graphite to several thousand degrees releasing atoms at high thermal velocities in a variety of atomic and ionic states. Only one of the atomic states can be selected for the subsequent RIS scheme. It is difficult to fully overlap the lasers with the fast-expanding plume of atoms and the RIS laser pulse must arrive typically within 10 μs of the ablation pulse before the expansion has developed too far. Krönert *et al.* [1] achieved an efficiency of one ion detected per 10^5 implanted in the graphite.

The proposal here is to use solid xenon at low temperature ($\sim$20K) as the catcher. The sample can be released as cold, free atoms simply by warming the catcher substrate to $\sim$80K. The release is slow and continuous compared to the RIS laser pulse structure (typically, a 10 ns pulse every 10 ms) so it is envisaged that the release would take place in the helium gas flow of an IGISOL-type ion source. This will slow the expansion of the atom cloud to allow a free-running pulsed laser system to ionize the sample. In this way, the method resembles that of the Leuven isotope separator laser ion source (IGLIS) [3]. Once the ions have been extracted, they may be mass-analysed and counted with conventional ion detectors.

2. Assessment of xenon layers with bismuth atom samples

2.1. APPARATUS FOR PREPARING THE XENON CATCHER

The apparatus used for the initial study of the solid xenon layers comprised a vacuum chamber built around the second stage (12 K) of a helium cryopump. A shield bolted to the first stage (70 K) extended to surround the second stage cold-head. A copper plate bolted to the cold-head was used as the substrate to hold the xenon layer. It was fitted with a thermometer and a heating wire to raise the temperature as required.

The xenon layers were formed by cooling the cryopump head down to its base temperature under high vacuum and then starting a slow flow of xenon gas down a metal capillary tube, which was directed at the copper substrate. A suitable layer could be formed in 2–3 min.

2.2. LASER FLUORESCENCE MEASUREMENTS

Bismuth was the element chosen for the assessment of the xenon layers. It is easy to produce as an atomic beam from ovens and the atom has a convenient transition from its ground state at 306.7 nm, which the Manchester group has already used in several studies of bismuth radioisotopes [4]. The laser fluorescence on this transition was used to monitor the deposition of the sample and the release of the bismuth atoms.

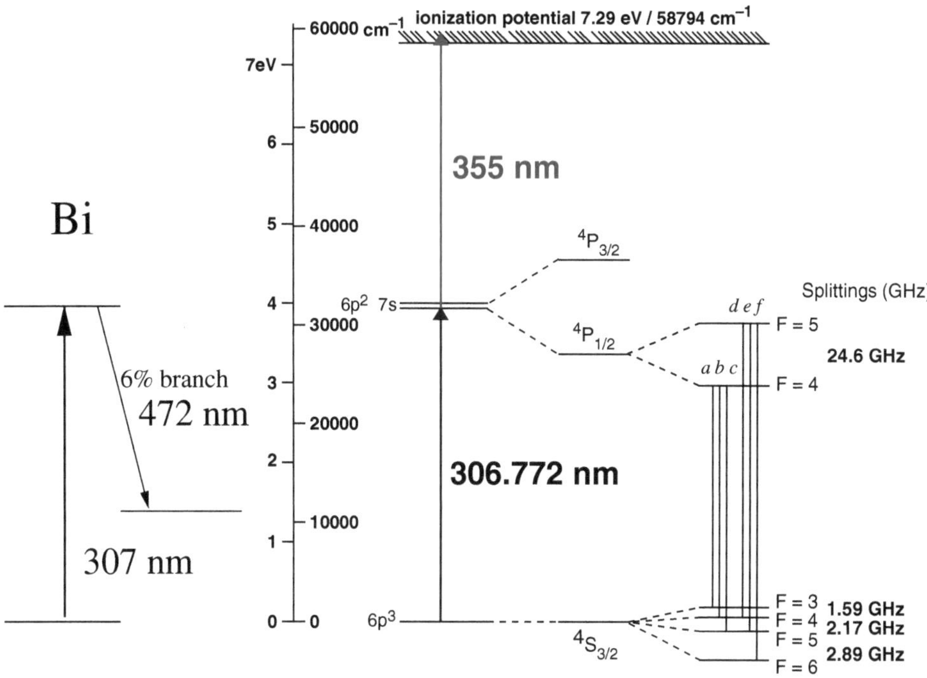

Figure 1. The atomic levels and hyperfine structure of the 307 nm transition in the stable isotope, ^{209}Bi.

Atom samples were produced from a small graphite tube oven mounted in a port of the vacuum chamber and directed at the xenon layer. The graphite was Joule-heated using a 10 V, 30 A DC supply. The bismuth atom sample was deposited in the xenon layer while it was formed.

Laser light to study the bismuth sample on the 307 nm line was produced by intra-cavity frequency-doubling in a lithium iodate crystal positioned at the auxiliary waist in a Spectra-Physics 380 D ring dye laser. The laser operated with Rhodamine 6 G dye and was pumped by a 5 W argon ion laser. The 0.3 mW laser beam was arranged to pass 5 mm above the xenon layer. Fluorescent light from this region was collected by a lens and imaged on a photomultiplier tube. The atomic structure and hyperfine structure of the 307 nm transition is shown in Figure 1. Narrow bandwidth filters were placed in front of the photomultiplier tube so that the laser-induced fluorescence could be observed on the 472 nm decay transition in the absence of Rayleigh-scattered and oven light.

The laser-fluorescence spectra in Figure 2 cover the hyperfine components labelled *d*, *e* and *f* in Figure 1. The upper spectrum was obtained for bismuth vapour in a low-pressure buffer gas near room temperature. The lower trace was obtained from a sample that was held in the solid xenon layer for 24 h before being released by applying a small amount of heating to the copper substrate. The half-widths of these resonances were 0.36 GHz compared with 0.80 GHz for the

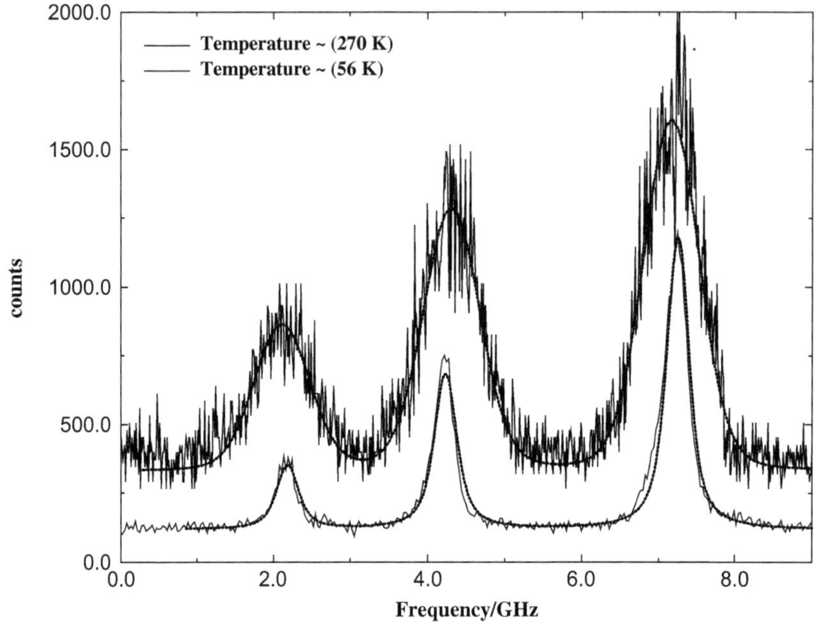

Figure 2. Laser fluorescence signals from bismuth vapour. The lower spectrum was obtained form a sample released from solid xenon after being held for 24 h.

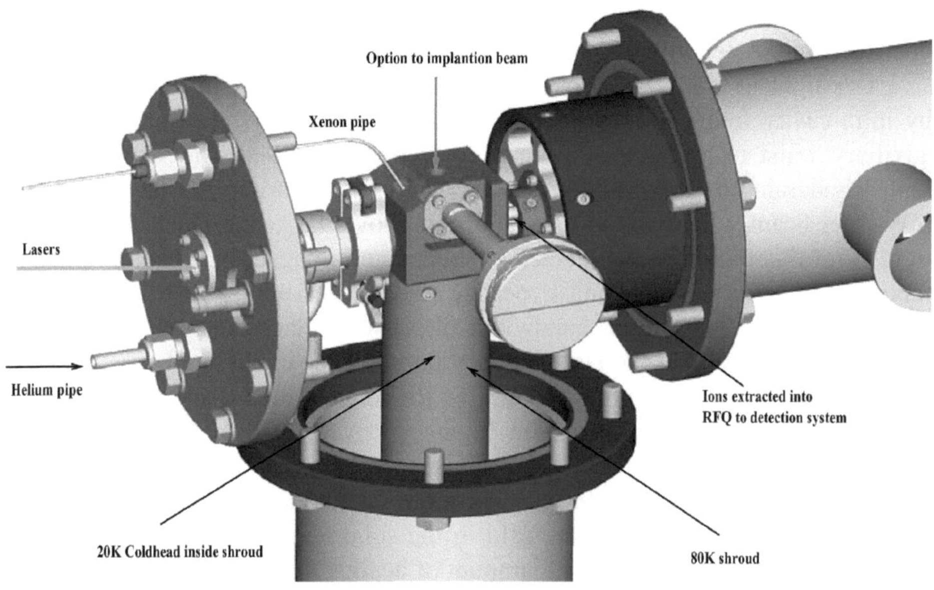

Figure 3. Design of the gas cell and the connection to the coldhead shroud.

fully thermalized bismuth sample near room temperature, and corresponds to a release temperature of 56 K. The entire sample could be released in typically 40 s.

2.3. FUTURE DEVELOPMENTS

New apparatus has been designed and built at Manchester to test the IGLIS method of detecting the released atoms (Figure 3). The gas cell is mounted on a shroud from the first stage (80 K) of the cold-head. The copper cold-finger from the second stage (20 K) extends into the gas cell volume and will hold the solid xenon layer. A port in the top side of the gas cell can be opened to allow an ion beam to be implanted into the layer.

Several resonance ionization schemes have been tested on bismuth. A simple two-step scheme using a tripled Nd:YAG laser for the second step is illustrated in Figure 1. The pulsed lasers enter the back of the gas cell. When the layer is heated for release of the sample, laser-induced ions will be swept through the gas cell nozzle into an RF quadrupole mass filter, which will deliver the ions to an electron multiplier for counting.

3. Application to isomers populated in α-decay

One application of the solid xenon catcher method is in the study of daughter nuclei produced by α-decay. For example, the ^{229m}Th isomer has attracted the attention of many nuclear and atomic groups because it lies only 3.5 ± 1.0 eV [5] from the nuclear ground state. It could be used to study new properties concerning the interaction of the nucleus with its atomic environment. It is proposed to deposit a thin open source of ^{233}U on the copper substrate, which is then covered with a solid xenon layer. Up to half of the ^{229,229m}Th products recoil out into the xenon where they accumulate. The substrate and xenon layer would be held in an (initially evacuated) IGLIS-type ion source [3]. Once the sample had been accumulated, the helium or argon gas flow would be started and the sample allowed to evaporate into it. A subsequent laser spectroscopic measurement on the acquired samples using the IGLIS method [3] could confirm the existence of the isomer through observation of its hyperfine structure. The measurement could also provide the static nuclear moments to confirm the state's structure. Once its detection is demonstrated, then its lifetime can be measured from the signal dependence on holding-time in the xenon layer. The isomer is apparently populated only weakly (a 2% branch) in the ^{233}U α-decay, but there are other sources such as ^{232}U, which could be used to optimize and test this method.

In summary, we have made xenon layers at 12 K, using the cold-head of a cryopump stage. We have stored bismuth atoms in solid xenon for up to 24 h without physical or chemical deterioration, and released them at around 80 K as free atoms. Laser-induced fluorescence has demonstrated the reduced

Doppler-broadening expected at these low temperatures. The project is on-going and will initially be applied to study daughter isotopes produced in α-decays of heavy elements.

References

1. Krönert U. *et al.*, *Appl. Phys. A* **44** (1987), 339.
2. Le Blanc F. *et al.*, *Phys. Rev. Lett.* **79** (1997), 2213.
3. Van Duppen P. *et al.*, *Hyperfine Interact.* **127** (2000), 401.
4. Billowes J. and Campbell P., *Hyperfine Interact.* **129** (2000), 289.
5. Helmer R. G. and Reich C. W., *Phys. Rev. C* **49** (1994), 1845.

Hyperfine Interactions (2005) 162:195–200
DOI 10.1007/s10751-005-9225-x

A High-Resolution Laser Setup for Determining Nuclear Moments

D. V. KARAIVANOV, S. G. ZEMLYANOI* and G. V. MYSHINSKY
Joint Institute for Nuclear Research, 141980 Dubna, Russia; e-mail: dvk@nrmail.jinr.ru

Abstract. A laser spectroscopic setup has been applied in the past to successfully investigate off-line the hyperfine structure splitting of radionuclides. In the present paper, its modification is described which will enable the determination of nuclear magnetic dipole and electric quadrupole moments with higher sensitivity and higher resolution.

Key Words: laser induced resonance fluorescence, off line experimental technique – sensitivity, selectivity, nuclear moments.

Our results on the quadrupole moment of ^{22}Na, the nuclear moments of ^{155}Eu, and the hyperfine anomaly of ^{151}Eu, ^{153}Eu, and ^{155}Eu have already been presented at the preceding Workshops in Poznan and also in [1, 2]. The experimental situation with these elements is very challenging due to the following reasons:

- The hyperfine splitting of the excited $3p\ ^2P_{3/2}$ -state of the investigated optical transition in Na is very small; as a consequence high resolution is required.
- High selectivity and accuracy are essential for determination of the hyperfine anomaly of the investigated Eu isotopes.
- Special temperature conditions for evaporating europium atoms are necessary.
- Radioactive contamination of the chamber, in which the atomic beam interacts with the laser light, has to be avoided.

Therefore the setup was modified leading to an essential improvement of its parameters and an extension of the range of nuclei which can be investigated.

An atomic-beam apparatus in the off-line mode has been employed. It is based on the detection of the laser-excited resonance fluorescence in a highly collimated atomic beam [3]. The well known orthogonal geometry is used to reduce the Doppler broadening. The experimental setup is presented in Figure 1. A cw-dye laser (Spectra Physics 380D pumped by an Ar ion laser) with active frequency stabilization was used for excitation. The spectra were recorded by

---

* Author for correspondence.

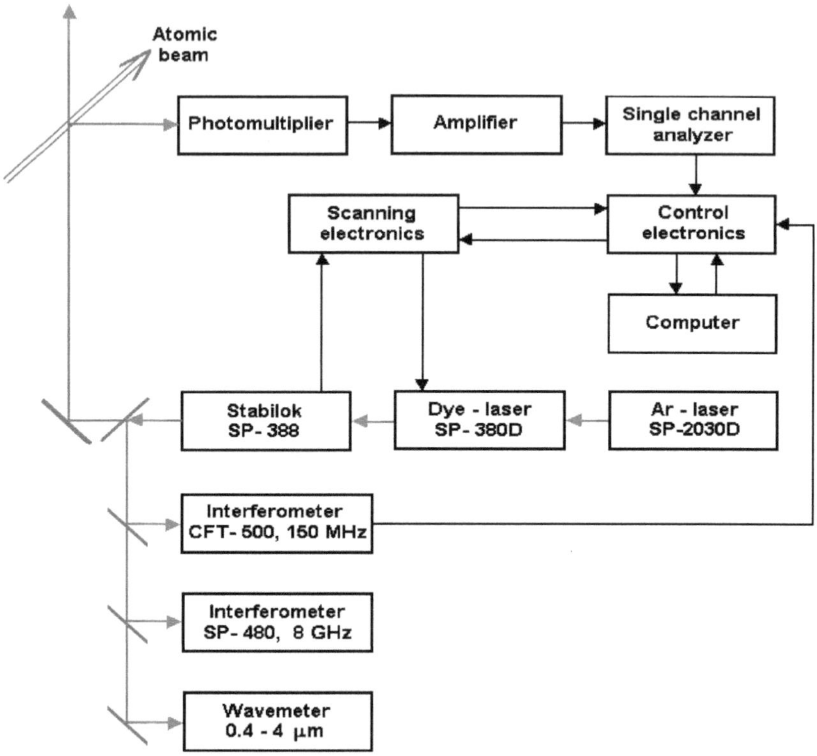

Figure 1. Scheme of the experimental setup.

means of an electronic system composed of an amplifier, a single-channel analyzer (operating as two-level discriminator) and control electronics. This system acts as a multi-channel analyser synchronized with the laser frequency tuning. A frequency calibration was provided by a temperature stabilized confocal interferometer with a free spectral range of 150 MHz. The data acquisition system as well as the experimental conditions was computer controlled. The scheme of the experimental setup and its detailed description can be found elsewhere [4, 5].

The atomization was performed in a tantalum crucible (Figure 2). Temperatures up to 1,500°C were obtained by an electrically heated tungsten oven. For obtaining higher temperatures electron bombardment was applied. The chamber is divided into two parts connected with a vacuum gate valve (Figure 3). The crucible and the oven are located in the upper part, and, in the lower part, the interaction zone is surrounded by the light-collecting system (Figure 4) and the photomultiplier. The light collection system geometrically covers a solid angle of $0.89 \cdot 4\pi$.

In the following, we describe four essential improvements of the experimental set-up and describe the achieved performance.

Increasing the temperature of evaporation and its stability. It has been observed that the temperature instabilities appearing at temperatures above

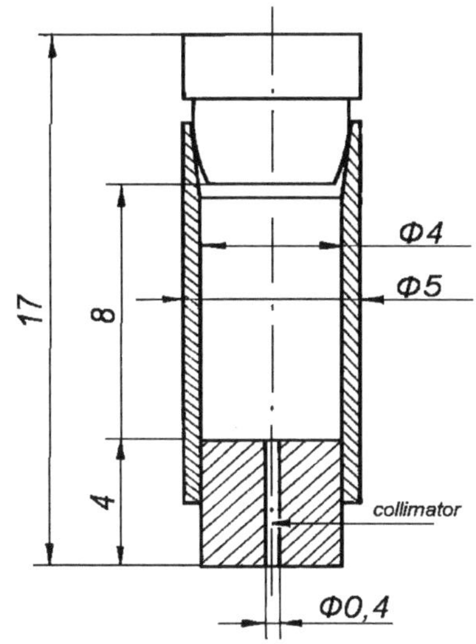

Figure 2. Crucible.

1,600°C were due to occasional local electrical breakdowns in the evaporation system. To cure these effects, significant changes of the evaporation system were undertaken:

- New, well conducting connectors for the tungsten wire have been incorporated allowing its rapid change and a good electrical contact up to the highest temperature applied.
- Oven size and position have been improved.
- Crucible and heater configurations were made sufficiently tight and in such a way that their exact coaxial positioning is assured at all temperatures applied.

Thus, we are able to keep temperatures up to 2,100°C with a high stability and over a long period of time. Moreover, the temperature can be both smoothly changed as well as steeply increased. The latter permits a dynamical control of the atomization rate: The temperature can be increased to the necessary high value only when the resonance lines of the investigated isotope are expected. Thus, the samples can be evaporated under optimal conditions.

Increasing the efficiency of the atomic beam collimation. This was achieved mainly by changing the dimensions of the crucible collimator. The ratio of the collimator length to its diameter was chosen to be 10 (see Figure 2), which is the optimal value predicted theoretically [6]. Thus, we could use more narrow diaphragms for atomic beam collimation while keeping the sensitivity of the setup unchanged. For example, in the case of sodium the diameter of the last diaphragm was 0.2 mm which corresponds to the Doppler broadening of 9.7 MHz which is less than the natural width of the sodium D_2 line.

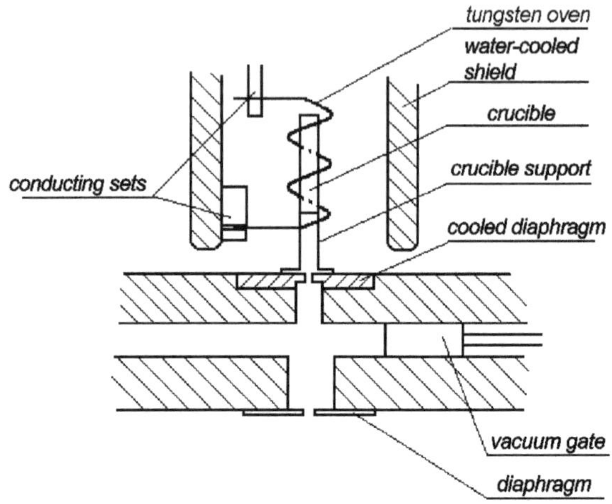

Figure 3. Evaporation system.

Increasing the signal-to-noise ratio. In our case, the main source of the background is black-body radiation from the heated crucible (see Figure 5) which causes 6,000 counts/s at a temperature of 2,100°C. The background by scattered laser light is substantially lower than the background due to heated crucible and is contributes typically 90 counts/s at a laser power of 600 μW. For reduction of background we used light filters placed in between the light-collecting system and the photomultiplier. Two different optical systems have been built up. One of them is used for an interference filter (Figure 4). The design of the chamber requires application of a concave lens. In this way, the entire solid angle of light collection is reduced to $0.61 \cdot 4\pi$ (see ω_1 and ω_2 angles in Figure 4). But this reduction is fully compensated by the increase of the signal-to-noise ratio by a factor of 30 at the maximal crucible temperature of 2,100°C. The efficiency of the photomultiplier (PM−136) is about 10% at 531 nm. The total efficiency of light detection system is of the order of 10^{-2} taking into account the transmission of the interference filter and reflection losses.

The second system employs usual coloured glass filters and results into a six times increased signal-to-noise ratio at crucible temperature of 2,100°C. The dependence of the background on the temperature is shown in Figure 5. In this case an interesting peculiarity has been observed: Some colour glass filters increase the background as a result of radioactive admixtures contained in the glass. For this reason, caution in using such filters is advisable.

Decreasing the fluctuations of the background. The background fluctuations depend mainly on the operating conditions of the photomultiplier [7]. This is especially true in the case when the multiplier operates in the single-photon counting mode, which needs highest gain and therefore maximal voltage. The most effective and simple way for decreasing the background fluctuations was found to be a highly precise voltage divider which was designed with great care. By this measure, the background fluctuation decreased more than two times.

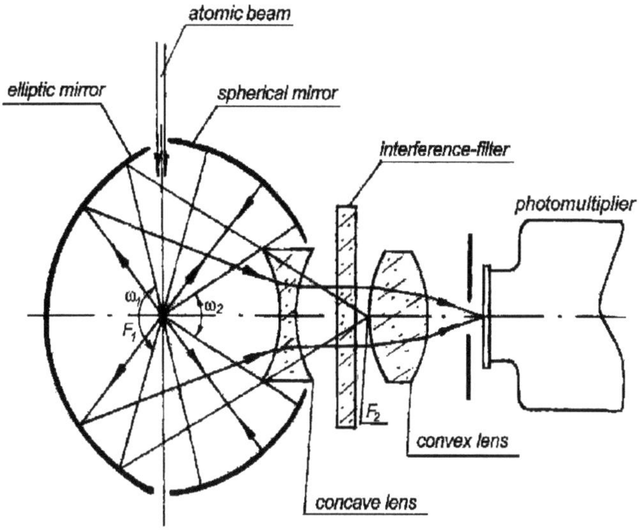

Figure 4. Light collecting system with interference filter.

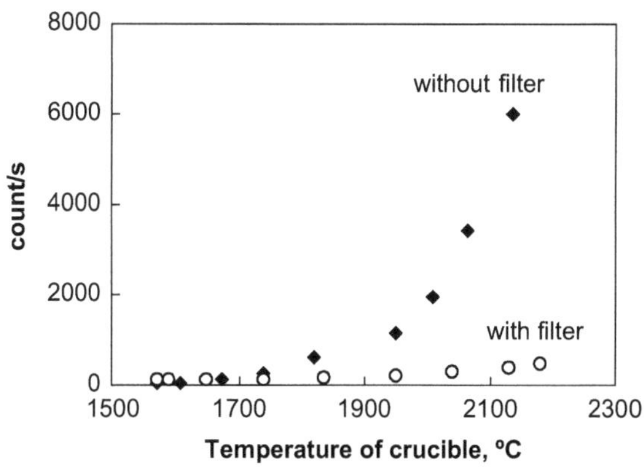

Figure 5. Dependence of the background signal on the temperature.

In conclusion, the following main results were obtained for the atomic beam setup: A sensitivity limit of the order of 10^{12} atoms in the sample was achieved with a collimation resulting in a resolution of the order of some megahertz. Generally, our experiments are carried out with samples contaminated by a large number of atoms of stable isotopes. Their signal can be two to three orders of magnitude larger than that of the isotope under investigation. This is an essential disadvantage, for example, in the case of the investigation of ^{22}Na in the D_2 transition [1], where the Lorentz wings of the ^{23}Na signal lead to an interfering fluorescence signal which is 10^2 times larger than the background.

Test experiments have shown that with mass-separated samples our sensitivity is of the order of 10^{11} atoms for the isotope under investigation. This is on the same level of sensitivity as reached by use of the off-line apparatus at Karlsruhe [8, 9] which has the best performance reported up to now.

The upcoming experiments planned to be performed in the Laboratory of Nuclear Reactions with the described off-line setup are the following: measurement of the electric quadrupole moment of ^{24}Na, study of the nuclear moments of ^{152}Eu, where our preliminary results on ^{152}Eu indicate evidence for a large hyperfine anomaly between ^{152}Eu and ^{151}Eu, and study of the nuclear moments and charge radii changes of long-lived transuranium isotopes in the region of the suggested neutron shell closure at $N = 152$.

References

1. Gangrsky Yu. P., Karaivanov D. V., Marinova K. P. *et al.*, *Eur. Phys. J.* **A 3** (1998), 313.
2. Gangrsky Yu. P., Karaivanov D. V., Markov B. N. *et al.*, *Eur. Phys. J., D* **11** (2000), 341.
3. Gangrsky Yu. P., Marinova K. P., Markov B. N. *et al.*, *Izv. AN USSR, Phys.* **ser. 49** (1985), 2261.
4. Gangrsky Yu. P., Zemlyanoi S. G., Kuldjanov B. K. *et al.*, *Opt. Spectrosc.* **81** (1989), 191.
5. Anastassov A., Gangrsky Yu. P., Marinova K. P. *et al.*, *Hyperfine Interact.* **77** (1992), 31.
6. Nadjakov E. G., Preprint JINR E6-86-231. 1986. 16 p.
7. Vetohin S. S., Gulakov I. R., Pertzev A. N., *Single-Electron Photodetectors. Atom Edition*, 1979, p. 198.
8. Nowicki G., Bekk K., Göring S. *et al.*, *Phys. Rev., C* **11** (1978), 2369.
9. Andl A., Bekk K., Göring S. *et al.*, *Phys. Rev., C* **26** (1982), 2194.

Hyperfine Interactions (2005) 162: 201–202

Author Index to Volume 162 (2005)

MIX
Papier aus verantwortungsvollen Quellen
Paper from responsible sources
FSC® C105338

If you have any concerns about our products,
you can contact us on
ProductSafety@springernature.com

In case Publisher is established outside the EU,
the EU authorized representative is:
Springer Nature Customer Service Center GmbH
Europaplatz 3, 69115 Heidelberg, Germany

Printed by Libri Plureos GmbH
in Hamburg, Germany